Current Topics in Microbiology and Immunology

Volume 373

Series Editors

Klaus Aktories
Medizinische Fakultät, Institut für Experimentelle und Klinische Pharmakologie und Toxikologie, Abt. I Albert-Ludwigs-Universität Freiburg, Albertstr. 25, 79104 Freiburg, Germany

Richard W. Compans
Department of Microbiology and Immunology, Emory University, 1518 Clifton Road, CNR 5005, Atlanta, GA 30322, USA

Max D. Cooper
Department of Pathology and Laboratory Medicine, Georgia Research Alliance, Emory University, 1462 Clifton Road, Atlanta, GA 30322, USA

Jorge E. Galan
Boyer Ctr. for Molecular Medicine, School of Medicine, Yale University, 295 Congress Avenue, room 343, New Haven, CT 06536-0812, USA

Yuri Y. Gleba
ICON Genetics AG, Biozentrum Halle, Weinbergweg 22, 06120 Halle, Germany

Tasuku Honjo
Faculty of Medicine, Department of Medical Chemistry, Kyoto University, Sakyo-ku, Yoshida, Kyoto 606-8501, Japan

Yoshihiro Kawaoka
School of Veterinary Medicine, University of Wisconsin-Madison, 2015 Linden Drive, Madison, WI 53706, USA

Bernard Malissen
Centre d'Immunologie de Marseille-Luminy, Parc Scientifique de Luminy, Case 906, 13288 Marseille Cedex 9, France

Fritz Melchers
Max Planck Institute for Infection Biology, Charitéplatz 1, 10117 Berlin, Germany

Michael B. A. Oldstone
Department of Immunology and Microbial Science, The Scripps Research Institute, 10550 North Torrey Pines Road, La Jolla, CA 92037, USA

Rino Rappuoli
Novartis Vaccines, Via Fiorentina 1, Siena 53100, Italy

Peter K. Vogt
Department of Molecular and Experimental Medicine, The Scripps Research Institute, 10550 North Torrey Pines Road, BCC-239, La Jolla, CA 92037, USA

Honorary Editor: Hilary Koprowski (deceased)
Biotechnology Foundation, Inc., Ardmore, PA, USA

For further volumes:
http://www.springer.com/series/82

Thomas Boehm · Yousuke Takahama
Editors

Thymic Development and Selection of T Lymphocytes

Responsible series editor: Tasuku Honjo

Editors

Thomas Boehm
Department of Developmental Immunology
Max-Planck-Institute of Immunobiology
  and Epigenetics
Freiburg
Germany

Yousuke Takahama
Division of Experimental Immunology
Institute for Genome Research
University of Tokushima
Tokushima
Japan

ISSN 0070-217X   ISSN 2196-9965 (electronic)
ISBN 978-3-642-40251-7   ISBN 978-3-642-40252-4 (eBook)
DOI 10.1007/978-3-642-40252-4
Springer Heidelberg New York Dordrecht London

Library of Congress Control Number: 2013953598

© Springer-Verlag Berlin Heidelberg 2014
This work is subject to copyright. All rights are reserved by the Publisher, whether the whole or part of the material is concerned, specifically the rights of translation, reprinting, reuse of illustrations, recitation, broadcasting, reproduction on microfilms or in any other physical way, and transmission or information storage and retrieval, electronic adaptation, computer software, or by similar or dissimilar methodology now known or hereafter developed. Exempted from this legal reservation are brief excerpts in connection with reviews or scholarly analysis or material supplied specifically for the purpose of being entered and executed on a computer system, for exclusive use by the purchaser of the work. Duplication of this publication or parts thereof is permitted only under the provisions of the Copyright Law of the Publisher's location, in its current version, and permission for use must always be obtained from Springer. Permissions for use may be obtained through RightsLink at the Copyright Clearance Center. Violations are liable to prosecution under the respective Copyright Law. The use of general descriptive names, registered names, trademarks, service marks, etc. in this publication does not imply, even in the absence of a specific statement, that such names are exempt from the relevant protective laws and regulations and therefore free for general use.
While the advice and information in this book are believed to be true and accurate at the date of publication, neither the authors nor the editors nor the publisher can accept any legal responsibility for any errors or omissions that may be made. The publisher makes no warranty, express or implied, with respect to the material contained herein.

Printed on acid-free paper

Springer is part of Springer Science+Business Media (www.springer.com)

Preface

The thymus is an evolutionarily ancient primary lymphoid organ common to all vertebrates in which T-cell development takes place. Remarkably, in jawed vertebrates the overall histological structure of the thymus has not changed over several 100 million years, as illustrated by the fact that the thymus of cartilaginous fishes already possesses an outer cortical and inner medullary region. This surprising stability of tissue organization underlies a conserved mechanism of T-cell differentiation that is a characteristic of the adaptive immune functions in all vertebrates.

The two major aspects of thymopoiesis, namely the development of the stromal microenvironment of the thymus and the development of T-cells, are addressed by the papers in this volume. Yousuke Takahama and Graham Anderson and their colleagues discuss cell biological and molecular aspects of epithelial differentiation in the cortical and medullary regions of the thymus. Although these two compartments are functionally interconnected, their properties are quite distinct as they support different stages of thymocyte development. Lo and Allen discuss the molecular basis of positive selection, the process by which the thymic microenvironment influences the formation of the repertoire of the T-cell receptors (TCRs) expressed on developing T-cells. Because TCRs can also exhibit self-reactivity, this property needs to be carefully controlled to avoid undesired autoimmunity; in their chapter, Maria Mouchess and Mark Anderson address the mechanisms by which the thymus imposes the essential central tolerance on T-cells. The last two chapters elaborate additional important aspects of haematopoietic cell differentiation: Zhang and Bhandoola discuss recent progress in understanding the many factors that regulate the homing of T-cell precursors to the thymic rudiment and also touch upon the question of which cell type(s) colonize the thymus; finally, Tanaka and Taniuchi discuss how genetic and epigenetic mechanisms regulate the decision between CD4 and CD8 lineage differentiation.

The contributions in this volume not only provide state-of-the-art overviews of the various aspects of thymopoiesis written by leading experts in the field, they also illustrate how far we have come in our understanding of thymus development and T-cell differentiation and how these observations might be translated into improved diagnosis and treatment of various immunodeficiency and autoimmune disorders disrupting thymus function.

Thomas Boehm
Yousuke Takahama

Contents

Development and Function of Cortical Thymic Epithelial Cells   1
Kensuke Takada, Izumi Ohigashi, Michiyuki Kasai, Hiroshi Nakase
and Yousuke Takahama

Mechanisms of Thymus Medulla Development and Function   19
Graham Anderson, Song Baik, Jennifer E. Cowan, Amanda M. Holland,
Nicholas I. McCarthy, Kyoko Nakamura, Sonia M. Parnell,
Andrea J. White, Peter J. L. Lane, Eric J. Jenkinson
and William E. Jenkinson

**Self-Peptides in TCR Repertoire Selection and Peripheral
T Cell Function**   49
Wan-Lin Lo and Paul M. Allen

Central Tolerance Induction   69
Maria L. Mouchess and Mark Anderson

Trafficking to the Thymus   87
Shirley L. Zhang and Avinash Bhandoola

**The CD4/CD8 Lineages: Central Decisions and Peripheral
Modifications for T Lymphocytes**   113
Hirokazu Tanaka and Ichiro Taniuchi

Index   131

Development and Function of Cortical Thymic Epithelial Cells

Kensuke Takada, Izumi Ohigashi, Michiyuki Kasai, Hiroshi Nakase and Yousuke Takahama

Abstract The thymic cortex provides a microenvironment that supports the generation and T cell antigen receptor (TCR)-mediated selection of $CD4^+CD8^+TCR\alpha\beta^+$ thymocytes. Cortical thymic epithelial cells (cTECs) are the essential component that forms the architecture of the thymic cortex and induces the generation as well as the selection of newly generated T cells. Here we summarize current knowledge on the development, function, and heterogeneity of cTECs, focusing on the expression and function of $\beta5t$, a cTEC-specific subunit of the thymoproteasome.

Contents

1	Introduction	2
2	Development of cTECs	2
	2.1 Thymus Organogenesis	2
	2.2 TEC Progenitor Cells and cTEC Development	3
	2.3 $\beta5t$ Expression in cTECs	4
	2.4 cTECs in Thymic Involution	4
3	Function of cTECs	5
	3.1 Generation of DP Thymocytes	5
	3.2 TCR-Mediated Selection of DP Thymocytes	7
4	Heterogeneity of cTECs	10
	4.1 Molecular Heterogeneity of cTECs	10
	4.2 Thymic Nurse Cells	11
5	Conclusions	11
	References	12

K. Takada · I. Ohigashi · M. Kasai · H. Nakase · Y. Takahama (✉)
Division of Experimental Immunology, Institute for Genome Research,
University of Tokushima, 3-18-15 Kuramoto, Tokushima 770-8503, Japan
e-mail: takahama@genome.tokushima-u.ac.jp

Current Topics in Microbiology and Immunology (2014) 373: 1–17
DOI: 10.1007/82_2013_322
© Springer-Verlag Berlin Heidelberg 2013
Published Online: 24 April 2013

1 Introduction

The thymus provides a microenvironment that is essential for T cell development and repertoire formation. The thymic cortex supports early T cell development and repertoire selection of developing thymocytes, whereas the thymic medulla supports the establishment of self-tolerance of newly generated T cells by attracting positively selected thymocytes for further negative selection and regulatory T cell development. The thymic cortex contains a network of cortical thymic epithelial cells (cTECs) and a high density of immature thymocytes. Macrophages and dendritic cells are also found in the thymic cortex. This chapter summarizes current knowledge on the development, function, and heterogeneity of cTECs, focusing on the expression and function of β5t, a recently identified cTEC-specific subunit of the thymoproteasome.

2 Development of cTECs

cTECs are epithelial cells that are localized in the cortex of the thymus. cTECs are derived from TEC progenitor cells that originate from the endodermal epithelium of the third pharyngeal pouch. As the thymoproteasome subunit β5t is expressed exclusively in cTECs, a better understanding of the mechanisms of the cTEC-specific expression of β5t will be useful for clarifying the molecular mechanisms of the development and regeneration of cTECs.

2.1 Thymus Organogenesis

The third pharyngeal pouch, which generates the thymus and the parathyroid gland, is formed by embryonic day (Ed) 9.5 in mouse. Transcription factors such as Hoxa3, Pax1/9, Eya1, Six1/4, and Tbx1 regulate the formation of pharyngeal arches and pouches, including the organogenesis of the thymus, and a defect in any of those transcription factors leads to abnormal organogenesis of the thymus (Manley and Capecchi 1995; Su et al. 2001; Jerome and Papaioannou 2001; Hetzer-Egger et al. 2002; Zou et al. 2006). In human, Tbx1 deficiency caused by the deletion of chromosome 22q11.2 in DiGeorge syndrome patients is often associated with congenital hypoplasia of the thymus (Merscher et al. 2001; Lindsay et al. 2001; Jerome and Papaioannou 2001).

The development of the third pharyngeal pouch is followed by the formation of the thymus-specific and parathyroid-specific primordia, which are characterized by the expression of transcription factors Foxn1 and Gcm2, respectively (Nehls et al. 1994; Gordon et al. 2001). Foxn1 mutations are associated with the *nude* phenotype that is characterized by severe thymic hypoplasia in mouse and human (Nehls et al. 1994, 1996; Adriani et al. 2004). The Foxn1-expressing thymus primordium, which is detectable as early as Ed 11.5 in mouse, contains progenitor cells that give rise to thymic epithelial cells (TECs) including cTECs and medullary TECs (mTECs) (Blackburn et al. 1996; Gill et al. 2002; Bennett el al. 2002). Upon ectopic transplantation of the endoderm containing the third pharyngeal pouch region under the kidney capsule of Foxn1-deficient *nude* mouse, the functional thymus including cTECs and mTECs is generated in the graft (Gordon et al. 2004). Thus, TECs, including cTECs and mTECs, are derived from Foxn1-expressing endodermal epithelial cells that are generated at the third pharyngeal pouch.

Following the development of the Foxn1-expressing thymus primordium, the thymus is seeded by hematopoietic cells that produce T cells. The Foxn1-dependent thymus primordium produces CCL25, a CCR9 ligand chemokine, whereas the Gcm2-dependent parathyroid primordium produces CCL21, a CCR7 ligand chemokine. The coordination between Foxn1-dependent thymic and Gcm2-dependent parathyroid primordial leads to the establishment of the CCL25/CCR9- and CCL21/CCR7-mediated chemokine guidance essential for prevascular fetal thymus colonization (Liu et al. 2006).

2.2 TEC Progenitor Cells and cTEC Development

Single-cell-based clonal analysis of TECs isolated from Ed 12 thymus primordium revealed that cTECs and mTECs are derived from common TEC progenitor cells (Rossi et al. 2006). The differentiation of bipotent TEC progenitor cells into cTECs and mTECs is postnatally detectable, suggesting that the postnatal thymus is maintained by a continuous supply of TECs from TEC progenitor cells (Bleul et al. 2006). However, the molecular and cellular characteristics of common TEC progenitor cells are unclear.

Regarding the development of mTECs, it has been reported that Aire-expressing mTECs, which are crucial for the establishment of self-tolerance in T cells, are differentiated from mTEC-specific progenitor cells, which are characterized by the expression of tight junction molecules claudin-3 and claudin-4 (Hamazaki et al. 2007). NF-κB transcription factors activated by thymocyte-derived signals through TNF superfamily receptors, including RANK, CD40, and lymphotoxin-β receptor, have been identified to be essential for mTEC development and medulla formation (van Ewijk et al. 1994; Burkly et al. 1995; Shinkura et al. 1999; Boehm et al. 2003; Akiyama et al. 2005, 2008; Rossi et al. 2007; Hikosaka et al. 2008). However, few studies have addressed the mechanisms of cTEC development. Embryonic TECs that express the cell surface molecule CD205, which is expressed by cTECs and dendritic cells but not mTECs, are detectable as early as Ed 12 in mouse and may represent an intermediate stage of developing TECs between common TEC progenitor cells and mature cTECs (Shakib et al. 2009). The role of thymocyte-derived signals in the promotion of cTEC development has also been documented (Holländer et al. 1995; Klug et al. 1998, 2002; Zamisch et al. 2005; Fiorini et al. 2008).

2.3 β5t Expression in cTECs

β5t is a cTEC-specific subunit of the thymoproteasome, and is essential for the positive selection of the majority of CD8$^+$ T cells (Murata et al. 2007; Nitta et al. 2010; Takahama et al. 2012). As far as we are aware of, β5t is expressed exclusively in cTECs (Murata et al. 2007; Mat Ripen et al. 2011; Takahama et al. 2012). No other molecules so far reported are exclusively expressed in TEC lineages. For example, Foxn1 is additionally expressed in the skin (Weiner et al. 2007; Hu et al. 2010), whereas Aire is additionally expressed during early development (Schaller et al. 2008; Nishikawa et al. 2010). Thus, a better understanding of the mechanisms of cTEC-specific β5t expression will be useful for clarifying the molecular mechanisms of cTEC development.

β5t is detectable in mouse embryonic thymus as early as Ed 12.5, approximately one day after the detection of Foxn1 expression in the thymus primordium (Mat Ripen et al. 2011). β5t-expressing cells in Ed 12.5 fetal thymus tend to localize at the ventral and outer region, which is distinct from the dorsal and inner region where mTEC progenitor cells that express claudin-3 and claudin-4 tend to localize (Mat Ripen et al. 2011). β5t expression in Ed 12.5 fetal thymus is not diminished in CCR7 and CCR9 double-deficient mouse (Mat Ripen et al. 2011), in which fetal thymus colonization by T-lymphoid progenitor cells is severely defective (Liu et al. 2006; Calderón and Boehm 2011), suggesting that thymus colonization by T-lymphoid progenitor cells and subsequent signals from developing thymocytes are not required for embryonic β5t expression. On the other hand, β5t-expressing cells in Ed 13.5 fetal thymus are not detectable in Foxn1-deficient *nude* mice (Mat Ripen et al. 2011), indicating that Foxn1 directly or indirectly contributes to the expression of β5t.

Nevertheless, β5t is dispensable for the development of cTECs and the formation of the thymic cortex (Murata et al. 2007; Nitta et al. 2010).

2.4 cTECs in Thymic Involution

After puberty, the thymus diminishes in size and the thymic parenchyma is replaced with adipose tissue. The thymic involution is associated with decreases in the numbers of cTECs and mTECs (Gray et al. 2006) and the decline in the number and diversity of T cells produced (Douek et al. 1998; Rudd et al. 2011). Whereas cTECs decrease with age, their regenerative potential was documented by the employment of CCX-CKR1-mediated diphtheria toxin receptor induced ablation of cTECs (Rode and Boehm 2012).

3 Function of cTECs

Lymphoid progenitor cells enter the postnatal thymus through the blood vessels, which tend to localize around the corticomedullary junction. Upon entry into the thymus, lymphoid progenitor cells, or immature $CD4^-CD8^-$ double-negative (DN) thymocytes, migrate toward the subcapsular region of the thymic cortex and are initiated to undergo proliferation and differentiation into T-lineage cells. DN thymocytes that succeed in the in-frame $TCR\beta$ V(D)J rearrangement express the pre-TCR complex and thereby differentiate into $CD4^+CD8^+$ double-positive (DP) thymocytes. This developmental process is termed β-selection. DP thymocytes are highly motile in the microenvironment of the thymic cortex and scan self-peptides available in the thymic cortex. Positively selected thymocytes are induced to survive and further differentiate into $CD4^+CD8^-$ or $CD4^-CD8^+$ single positive (SP) thymocytes, which migrate into the thymic medulla for further development and selection before the export of T cells to the circulation. cTECs play a major role in providing the microenvironment for the generation of DP thymocytes and their selection in the thymic cortex.

3.1 Generation of DP Thymocytes

Immature DN thymocytes, which are often subdivided into four subpopulations based on the expression of CD44 and CD25, undergo proliferation and differentiation into DP thymocytes by progression through DN1 ($CD44^+CD25^-$), DN2 ($CD44^+CD25^+$), DN3 ($CD44^-CD25^+$), and DN4 ($CD44^-CD25^-$) stages. Signals via Notch, IL-7 receptor, and pre-TCR essentially support the generation of DP thymocytes (Thompson and Zúñiga-Pflücker 2011). The plasma membrane expression of the pre-TCR complex that contains successfully rearranged $TCR\beta$ and invariant pre-$TCR\alpha$ chains can induce an autonomous signal without any ligand engagement (Yamasaki and Saito 2007). However, the ligands for Notch and IL-7 receptor are provided by cTECs (Fig. 1).

3.1.1 DLL4

In mammals, the Notch family is composed of four transmembrane proteins; Notch1, Notch2, Notch3, and Notch4. The engagement of those Notch receptors by their ligands, including DLL1, DLL3, DLL4, Jagged1, and Jagged2, triggers the proteolytic cleavage of Notch proteins and generates the intracellular domain of Notch, which relocates into the nucleus and controls the transcription of target genes by the formation of a molecular complex with DNA-binding transcription factors (Radtke et al. 2004). In the thymus, various Notch receptors and ligands are expressed on both developing thymocytes and stromal cells (Radtke et al. 2004).

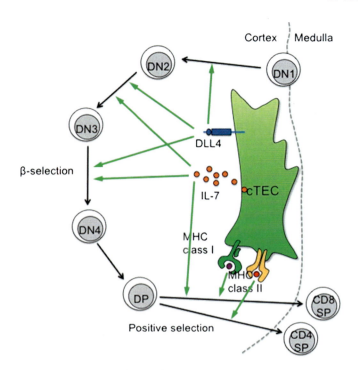

Fig. 1 A model for cTEC contributions to T cell development. The progression of proliferation and differentiation of DN thymocytes requires signals through Notch and IL-7 receptor at multiple stages. cTECs provide DLL4, the Notch ligand, and IL-7, the IL-7 receptor ligand. In addition, self-peptides presented on the surface of cTECs in association with MHC molecules essentially induce the positive selection of DP thymocytes. IL-7 is also involved in the development of CD8 SP thymocytes during the positive selection of thymocytes. The positively selected thymocytes differentiate into SP thymocytes and migrate to the thymic medulla, which provides the microenvironment that supports further thymocyte development and the establishment of self-tolerance in T cells

Notch1 signal continuously affects T cell development throughout the DN stages, including T cell lineage specification (Radtke et al. 1999; Wilson et al. 2001; Bell and Bhandoola 2008; Wada et al. 2008) and subsequent β-selection (Wolfer et al. 2002; Ciofani and Zúñiga-Pflücker 2005; Maillard et al. 2006). Although both DLL1 and DLL4 can induce thymocyte development from early precursors with multipotency in in vitro culture with OP9 stromal cells (Schmitt and Zúñiga-Pflücker 2002; Hozumi et al. 2004), DLL4 but not DLL1 expressed by cTECs plays an essential role in the induction of early thymocyte development in vivo (Hozumi et al. 2004, 2008; Koch et al. 2008). DLL4 is expressed by most cTECs whereas its expression in mTECs is limited to a small fraction and the loss of DLL4 in TECs leads to a complete block of T cell development and the ectopic appearance of B cells in the thymus (Koch et al. 2008; Hozumi et al. 2008).

3.1.2 IL-7

IL-7 signaling is also essential for early thymocyte development. The genetic ablation of IL-7 or IL-7 receptor results in severe defects in T cell development, which are associated with increased apoptosis and reduced cell cycle progression of DN thymocytes (Peschon et al. 1994; von Freeden-Jeffry et al. 1997). The impaired T cell development in the absence of the IL-7 signal is at least partly restored by protecting cells from apoptosis (Maraskovsky et al. 1997; Kondo et al. 1997; Pellegrini et al. 2004), suggesting that IL-7 plays an important role in providing survival signals for immature thymocytes. In addition to its important role during DN stages, IL-7 contributes to the specification of the differentiation of positively selected thymocytes into $CD4^-CD8^+$ SP thymocytes (Park et al. 2010). Studies of several lines of reporter mice harboring fluorescent protein genes controlled by IL-7 promoter elements have shown that IL-7 in the thymus is expressed in various stromal cells including cTECs, mTECs, and mesenchymal cells (Alves et al. 2009; Mazzucchelli et al. 2009; Hara et al. 2012). cTECs express high levels of IL-7 compared to other stromal cell populations, suggesting that cTECs play a major role in providing IL-7 to DN and DP thymocytes (Hara et al. 2012).

3.1.3 TGF-β

TGF-β is a cytokine that regulates the proliferation and differentiation of various cells. It was shown that TGF-β1 and TGF-β2 were expressed by cTECs, including the subcapsular epithelium, and regulated the generation of DP thymocytes through control of the cell cycle progression of $CD8^{low}$ immediate precursor cells (Takahama et al. 1994).

3.2 TCR-Mediated Selection of DP Thymocytes

DP thymocytes, which are newly generated in the thymic cortex, express a virgin repertoire of TCR specificities that are generated by the V(D)J rearrangement of TCR $\alpha\beta$ genes. DP thymocytes that interact with antigen-presenting cells through the low-affinity engagement of TCR with self-peptide-MHC complexes available in the thymic cortex are positively selected to survive and undergo further differentiation into $CD4^+CD8^-$ or $CD4^-CD8^+$ SP thymocytes. Within the thymic cortex, cTECs play a primary role as self-antigen-presenting cells that induce positive selection (Laufer et al. 1996) (Fig. 1). It is now known that intracellular proteolytic enzymes that are uniquely or highly expressed in cTECs contribute to the production of self-peptides that induce positive selection.

3.2.1 CD4 T Cell Lineage: Cathepsin L and TSSP

MHC class II molecules are loaded with peptides that are present in late endosomes. Lysosomal proteases degrade invariant (Ii) chain leaving MHC class II associated Ii peptide (CLIP) in the peptide-binding groove of MHC class II molecules (Neefjes et al. 2011). Distinct from other professional antigen-presenting cells, such as dendritic cells and mTECs, which highly express cathepsin S, cTECs express a unique set of lysosomal enzymes, including cathepsin L (Nakagawa et al. 1998; Honey et al. 2002) and thymus-specific serine protease (TSSP, Prss16) (Bowlus et al. 1999; Carrier et al. 1999; Gommeaux et al. 2009; Viret et al. 2011a). Cathepsin L mediates the degradation of Ii and the generation of self-peptides (Nakagawa et al. 1998). Disruption of the gene encoding cathepsin L in mice results in the reduction of CD4$^+$CD8$^-$ SP thymocytes and peripheral CD4$^+$ T cells by 60–80 %, despite the fact that the expression of surface MHC class II molecules by cTECs is unaffected (Nakagawa et al. 1998; Honey et al. 2002). In mice deficient in TSSP, the positive selection of T cells that express MHC class II restricted transgenic TCRs is impaired, whereas MHC class I restricted TCR-transgenic T cells are not affected (Gommeaux et al. 2009). The number of polyclonal CD4$^+$ T cells is not diminished in TSSP-deficient mice (Cheunsuk et al. 2005; Gommeaux et al. 2009). However, CD4$^+$ T cells generated in TSSP-deficient mice poorly respond to an antigenic peptide, and this poor response is correlated with the significant alteration of the dominant TCR-β chain repertoire expressed by antigen-specific CD4$^+$ T cells (Viret et al. 2011a). The development of CD4$^+$ T cells specific for an islet-derived self-antigen is also impaired in TSSP-deficient mice, and this impairment may be associated with the resistance of TSSP-deficient NOD mice to diabetes (Viret et al. 2011b). Thus, both cathepsin L and TSSP may contribute to the production of MHC class II associated self-peptides that are uniquely expressed by cTECs and induce the positive selection of CD4$^+$ T cells. The fact that the phenotypes of CD4$^+$ T cells in cathepsin L-deficient mice and TSSP-deficient mice are different suggests the non-overlapping roles of these two enzymes in cTECs, possibly in the generation of different sets of self-peptides.

3.2.2 Autophagy

Autophagy is a protein degradation process in that organelles and cytoplasmic proteins are digested via the lysosomes. Autophagy is activated in response to nutrient starvation and contributes to the recycling of unnecessary proteins into proteins essential for cell survival. It has been shown that autophagy, more specifically macroautophagy, is constitutively active in the majority of cTECs in normal mice even without nutrient deprivation, and that the positive selection of certain specificities of MHC class II restricted CD4$^+$ T cells is impaired in mice deficient in autophagy (Nedjic et al. 2008). Thus, probably by contributing to the supply of lysosome-derived peptides that are loaded into MHC class II molecules in cTECs, autophagy in cTECs is important for the positive selection of CD4$^+$ T cells.

3.2.3 CD8 T Cell Lineage: β5t-Containing Thymoproteasomes

MHC class I molecules present peptides that are derived from the cytoplasm. Proteasomes are cytoplasmic proteolytic complexes that play a pivotal role in the generation of peptides loaded on MHC class I molecules (Rock and Goldberg 1999). The proteolytic activity of proteasomes is mediated by three β subunits; β1, β2, and β5. Whereas the constitutively expressed form of proteasomes contains β1, β2, and β5 proteolytic subunits, IFNγ-stimulated cells and professional antigen-presenting cells, including dendritic cells and mTECs, produce a different set of proteolytic subunits, β1i, β2i, and β5i, forming an alternative form of proteasomes termed immunoproteasomes. Immunoproteasomes exhibit elevated chymotrypsin-like activity compared to the constitutively expressed form of proteasomes, preferentially generating cytoplasmic peptides that carry the hydrophobic carboxyl termini, which efficiently bind to MHC class I molecules and thereby contribute to efficient viral antigen presentation in virus-infected cells (Murata et al. 2009). A few years ago, an additional proteasome subunit termed β5t was identified (Murata et al. 2007). β5t is specifically expressed in cTECs and incorporated with β1i and β2i to form a novel form of proteasomes termed thymoproteasomes (Murata et al. 2007).

The diminished chymotrypsin-like activity of thymoproteasomes compared to those of the other two forms of proteasomes, i.e., constitutively expressed proteasomes and immunoproteasomes, results in reduced efficiency to produce hydrophobic carboxyl terminal anchor residues for MHC class I binding (Murata et al. 2007; Florea et al. 2010). Hence, cTECs are likely to present a unique repertoire of MHC class I associated self-peptides that are distinct from those expressed elsewhere in the body.

In β5t-deficient mice, the positive selection of CD8 SP thymocytes is impaired but the positive selection of CD4 SP thymocytes is unaffected. The cellularity of CD8 SP thymocytes and peripheral CD8$^+$ T cells in β5t-deficient mice is reduced to approximately 20–30 % of that in normal mice (Murata et al. 2007; Nitta et al. 2010). The remaining cells in β5t-deficient mice are likely selected by an altered set of MHC class I associated self-peptides expressed by cTECs, where β5i-containing immunoproteasomes are compensatively produced (Nitta et al. 2010; Takahama et al. 2012). The magnitude of the dependence on β5t for positive selection is different among different TCRs: for instance, the development of HY TCR and P14 TCR-transgenic CD8$^+$ T cells is severely affected, whereas the development of 2C TCR and OT-I TCR-transgenic CD8$^+$ T cells is only moderately affected in β5t-deficient mice (Nitta et al. 2010). In addition, β5t-deficient mice lack reactivity to some, but not all, allogeneic antigens and viral infections (Nitta et al. 2010). Those findings indicate that CD8$^+$ T cells generated in the absence of thymoproteasomes bear an altered TCR repertoire and that β5t-containing thymoproteasomes exclusively expressed by cTECs are essential for the positive selection of the majority of CD8$^+$ T cells that carry functional competence.

4 Heterogeneity of cTECs

The thymic cortex in the postnatal thymus contains several distinct microenvironments including the subcapsular region, the central cortex, and the peri-medullary cortex (Boyd et al. 1993; Griffith et al. 2009). Staining profiles with monoclonal antibodies and microscopic morphology as well as the location in the thymic cortex has revealed at least four different subtypes in cTECs (Boyd et al. 1993). However, the functional difference among these four cTEC subtypes is unclear. Global gene expression analysis of the subcapsular region, the central cortex, and the peri-medullary cortex has shown inequality in the gene expression profiles of those regions (Griffith et al. 2009).

4.1 Molecular Heterogeneity of cTECs

As described above (see Sect. 3), the key functions of cTECs are mediated by a combination of molecules such as DLL4, IL7, MHC, β5t, cathepsin L, and TSSP. It is therefore important to address whether cTECs consist of heterogeneous populations that express a fraction of those molecules or represent a homogeneous population that expresses all of those molecules. As the thymic cortex provides at least two major functionally different microenvironments, namely, one for the generation of DP thymocytes and another for the TCR-mediated selection of DP thymocytes, it is tempting to speculate that those two different cortical microenvironments may be mediated by functionally distinct cTEC subpopulations. However, flow cytometric analysis has shown that the majority (>80 %) of CD249 (Ly51, BP1)$^+$ cTECs in the postnatal thymus express both DLL4 and β5t (Koch et al. 2008; Nakagawa et al. 2012). The majority of cTECs also produce IL-7 (Hara et al. 2012). The results suggest that DLL4-expressing cTECs and IL-7-expressing cTECs, which contribute to the induction of early T cell development, largely overlap with β5t-expressing cTECs, which contribute to the TCR-mediated selection of cortical thymocytes.

Nonetheless, cTECs in the postnatal thymus consist of heterogeneous subpopulations. Probably the most widely recognized heterogeneity in cTECs is the variety in the expression intensity of class II MHC molecules (Yang et al. 2006; Shakib et al. 2009). It can be ontogenically presumed that class II MHC^{low} cTECs may represent immature cells that later develop into class II MHC^{high} mature cTECs, although this presumption has not been formally evaluated. cTECs are also heterogeneous in the intensity of CD205 expression, consisting of CD205^{low} cTECs and CD205^{high} cTECs (Shakib et al. 2009; Mat Ripen et al. 2011). The heterogeneity of cTECs in the expression intensity of other molecules, including DLL4 and IL7, has also been noted (Koch et al. 2008; Hara et al. 2012).

4.2 Thymic Nurse Cells

Regarding the heterogeneity in cTECs, it should be interesting to discuss the thymic nurse cell (TNC). TNC is a large epithelial cell that completely envelops many viable lymphoid cells within its intracellular vesicles and is isolated by the protease digestion of mouse and rat thymus tissues (Wekerle and Ketelson 1980; Wekerle et al. 1980; Ritter et al. 1981; Kyewski and Kaplan 1982). TNC complexes are found not only in rodent but also in many vertebrate species, including human, bird, and fish (Ritter et al. 1981; van de Wijngaert et al. 1983; Rieker et al. 1995; Flaño et al. 1996). It was hypothesized that TNCs provide a microenvironment that is necessary for thymocyte proliferation and differentiation and that the intra-TNC differentiation is an essential step in intrathymic T cell development (Wekerle and Ketelson 1980; Wekerle et al. 1980; Shortman et al. 1986; de Waal Malefijt et al. 1986). It was further hypothesized that the TNC complex is a site for the positive and negative selection of T cells (Wick et al. 1991; Aguilar et al. 1994; Guyden and Pezzano 2003). However, how TNCs are involved in T cell development and selection had been unclear until recently. It had been even questioned whether the TNC complexes indeed represent structures that are present in the thymus in vivo or are artificially generated during cell isolation procedures in vitro (Kyewski and Kaplan 1982; Tousaint-Demylle et al. 1990; Pezzano et al. 2001). We have recently shown that cTECs, but not mTECs, in the postnatal mouse thymus contain cellular complexes that are tightly associated with many thymocytes. Approximately 10 % of β5t-expressing cTECs resemble TNC complexes that completely enclose CD4$^+$CD8$^+$ cortical thymocytes. The cTEC-thymocyte complexes, including TNCs, are detected in the thymic cortex intravitally. Interestingly, the TNC complexes appear late during ontogeny but are not detected in the adult thymus of various TCR-transgenic mouse lines in that the majority of thymocytes can be positively selected, indicating that the formation of the TNC complex is not an absolute requirement for T cell differentiation, including the process of positive selection. Instead, the TNC complex represents a persistent interaction between adhesive cTECs and long-lived DP thymocytes that undergo secondary TCRα rearrangement. Thus, TNCs represent a subpopulation of β5t$^+$ cTECs that provide a microenvironment for the optimization of TCR selection by supporting the secondary TCR-Vα rearrangement in long-lived DP thymocytes (Nakagawa et al. 2012).

5 Conclusions

We have summarized current knowledge on the development, function, and heterogeneity of cTECs, focusing on the expression and function of β5t. Despite the knowledge that cTECs belong to endodermal epithelial cells and are neither lymphoid nor hematopoietic cells, cTECs are the essential component of the adaptive immune system by expressing molecules, including β5t and MHC class II, whose functions are primarily appreciated in the adaptive immune system and whose expression is detectable only in vertebrates, which possess the adaptive immune system (Boehm 2009; Takahama et al. 2010). Thus, the acquisition of a profound understanding of the immune system in vertebrates, including human, requires a better understanding of the biology of non-hematopoietic cells, including cTECs and mTECs, in addition to a better understanding of the biology of hematopoietic cells, including lymphoid cells.

Acknowledgments This work was supported by Grants-in-Aid for Scientific Research from MEXT and JSPS (grant numbers 23249025 and 24111004).

References

Adriani M, Martinez-Mir A, Fusco F, Busiello R, Frank J, Telese S, Matrecano E, Ursini MV, Christiano AM, Pignata C (2004) Ancestral founder mutation of the nude (FOXN1) gene in congenital severe combined immunodeficiency associated with alopecia in southern Italy population. Ann Hum Genet 68:265–268

Aguilar LK, Agilar-Cordova E, Cartwright J Jr, Belmont JW (1994) Thymic nurse cells are sites of thymocyte apoptosis. J Immunol 152:2645–2651

Akiyama T, Maeda S, Yamane S, Ogino K, Kasai M, Kajiura F, Matsumoto M, Inoue J (2005) Dependence of self-tolerance on TRAF6-directed development of thymic stroma. Science 308:248–251

Akiyama T, Shimo Y, Yanai H, Qin J, Ohshima D, Maruyama Y, Asaumi Y, Kitazawa J, Takayanagi H, Penninger JM, Matsumoto M, Nitta T, Takahama Y, Inoue J (2008) The tumor necrosis factor family receptors RANK and CD40 cooperatively establish the thymic medullary microenvironment and self-tolerance. Immunity 29:423–437

Alves NL, Huntington ND, Rodewald HR, Di Santo JP (2009) Thymic epithelial cells: the multi-tasking framework of the T cell "cradle". Trends Immunol 30:468–474

Bell JJ, Bhandoola A (2008) The earliest thymic progenitors for T cells possess myeloid lineage potential. Nature 452:764–767

Bennett AR, Farley A, Blair NF, Gordon J, Sharp L, Blackburn CC (2002) Identification and characterization of thymic epithelial progenitor cells. Immunity 16:803–814

Blackburn CC, Augustine CL, Li R, Harvey RP, Malin MA, Boyd RL, Miller JF, Morahan G (1996) The nu gene acts cell-autonomously and is required for differentiation of thymic epithelial progenitors. Proc Natl Acad Sci USA 93:5742–5746

Bleul CC, Carbeaux T, Reuter A, Fisch P, Monting JS, Boehm T (2006) Formation of a functional thymus initiated by a postnatal epithelial progenitor cell. Nature 441:992–996

Boehm T, Scheu S, Pfeffer K, Bleul CC (2003) Thymic medullary epithelial cell differentiation, thymocyte emigration, and the control of autoimmunity require lympho-epithelial cross talk via LTβR. J Exp Med 198:757–769

Boehm T (2009) The adaptive phenotype of cortical thymic epithelial cells. Eur J Immunol 39:944–947

Bowlus CL, Ahn J, Chu T, Gruen JR (1999) Cloning of a novel MHC-encoded serine peptidase highly expressed by cortical epithelial cells of the thymus. Cell Immunol 196:80–86

Boyd RL, Tucek CL, Godfrey DI, Izon DJ, Wilson TJ, Davidson NJ, Bean AG, Ladyman HM, Ritter MA, Hugo P (1993) The thymic microenvironment. Immunol Today 14:445–459

Development and Function of Cortical Thymic Epithelial Cells 13

Burkly L, Hession C, Ogata L, Reilly C, Marconi LA, Olson D, Tizard R, Cate R, Lo D (1995) Expression of relB is required for the development of thymic medulla and dendritic cells. Nature 373:531–536

Calderón L, Boehm T (2011) Three chemokine receptors cooperatively regulate homing of hematopoietic progenitors to the embryonic mouse thymus. Proc Natl Acad Sci USA 108:7517–7522

Carrier A, Nguyen C, Victorero G, Granjeaud S, Rocha D, Bernard K, Miazek A, Ferrier P, Malissen M, Naquet P, Malissen B, Jordan BR (1999) Differential gene expression in CD3epsilon- and RAG1-deficient thymuses: definition of a set of genes potentially involved in thymocyte maturation. Immunogenetics 50:255–270

Cheunsuk S, Lian ZX, Yang GX, Gershwin ME, Gruen JR, Bowlus CL (2005) Prss16 is not required for T-cell development. Mol Cell Biol 25:789–796

Ciofani M, Zúñiga-Pflücker JC (2005) Notch promotes survival of pre-T cells at the beta-selection checkpoint by regulating cellular metabolism. Nat Immunol 6:881–888

de Waal Malefijt R, Leene W, Roholl PJ, Wormmeester J, Hoeben KA (1986) T cell differentiation within thymic nurse cells. Lab Invest 55:25–34

Douek DC, McFarland RD, Keiser PH, Gage EA, Massey JM, Haynes BF, Polis MA, Haase AT, Feinberg MB, Sullivan JL, Jamieson BD, Zack JA, Picker LJ, Koup RA (1998) Changes in thymic function with age and during the treatment of HIV infection. Nature 396:690–695

Fiorini E, Ferrero I, Merck E, Favre S, Pierres M, Luther SA, MacDonald HR (2008) Thymic crosstalk regulates delta-like 4 expression on cortical epithelial cells. J Immunol 181:8199–8203

Flaño E, Alvarez F, López-Fierro P, Razquin BE, Villena AJ, Zapata AG (1996) In vitro and in situ characterization of fish thymic nurse cells. Dev Immunol 5:17–24

Florea BI, Verdoes M, Li N, van der Linden WA, Geurink PP, van den Elst H, Hofmann T, de Ru A, van Veelen PA, Tanaka K, Sasaki K, Murata S, den Dulk H, Brouwer J, Ossendorp FA, Kisselev AF, Overkleeft HS (2010) Activity-based profiling reveals reactivity of the murine thymoproteasome-specific subunit $\beta 5t$. Chem Biol 17:795–801

Gill J, Malin M, Holländer GA, Boyd R (2002) Generation of a complete thymic microenvironment by MTS24+ thymic epithelial cells. Nat Immunol 3:635–642

Gommeaux J, Grégoire C, Nguessan P, Richelme M, Malissen M, Guerder S, Malissen B, Carrier A (2009) Thymus-specific serine protease regulates positive selection of a subset of CD4+ thymocytes. Eur J Immunol 39:956–964

Gordon J, Bennett AR, Blackburn CC, Manley NR (2001) Gcm2 and Foxn1 mark early parathyroid- and thymus-specific domains in the developing third pharyngeal pouch. Mech Dev 103:141–143

Gordon J, Wilson VA, Blair NF, Sheridan J, Farley A, Wilson L, Manley NR, Blackburn CC (2004) Functional evidence for a single endodermal origin for the thymic epithelium. Nat Immunol 5:546–553

Gray DH, Seach N, Ueno T, Milton MK, Liston A, Lew AM, Goodnow CC, Boyd RL (2006) Developmental kinetics, turnover, and stimulatory capacity of thymic epithelial cells. Blood 108:3777–3785

Griffith AV, Fallahi M, Nakase H, Gosink M, Young B, Petrie HT (2009) Spatial mapping of thymic stromal microenvironments reveals unique features influencing T lymphoid differentiation. Immunity 31:999–1009

Guyden JC, Pezzano M (2003) Thymic nurse cells: a microenvironment for thymocyte development and selection. Int Rev Cytol 223:1–37

Hamazaki Y, Fujita H, Kobayashi T, Choi Y, Scott HS, Matsumoto M, Minato N (2007) Medullary thymic epithelial cells expressing Aire represent a unique lineage derived from cells expressing claudin. Nat Immunol 8:304–311

Hara T, Shitara S, Imai K, Miyachi H, Kitano S, Yao H, Tani-ichi S, Ikuta K (2012) Identification of IL-7-producing cells in primary and secondary lymphoid organs using IL-7-GFP knock-in mice. J Immunol 189:1577–1584

Hetzer-Egger C, Schorpp M, Haas-Assenbaum A, Balling R, Peters H, Boehm T (2002) Thymopoiesis requires Pax9 function in thymic epithelial cells. Eur J Immunol 32:1175–1181

Hikosaka Y, Nitta T, Ohigashi I, Yano K, Ishimaru N, Hayashi Y, Matsumoto M, Matsuo K, Penninger JM, Takayanagi H, Yokota Y, Yamada H, Yoshikai Y, Inoue J, Akiyama T, Takahama Y (2008) The cytokine RANKL produced by positively selected thymocytes fosters medullary thymic epithelial cells that express autoimmune regulator. Immunity 29:438–450

Holländer GA, Wang B, Nichogiannopoulou A, Platenburg PP, van Ewijk W, Burakoff SJ, Gutierrez-Ramos JC, Terhorst C (1995) Developmental control point in induction of thymic cortex regulated by a subpopulation of prothymocytes. Nature 373:350–353

Honey K, Nakagawa T, Peters C, Rudensky A (2002) Cathepsin L regulates $CD4^+$ T cell selection independently of its effect on invariant chain: a role in the generation of positively selecting peptide ligands. J Exp Med 195:1349–1358

Hozumi K, Negishi N, Suzuki D, Abe N, Sotomaru Y, Tamaoki N, Mailhos C, Ish-Horowicz D, Habu S, Owen MJ (2004) Delta-like 1 is necessary for the generation of marginal zone B cells but not T cells in vivo. Nat Immunol 5:638–644

Hozumi K, Mailhos C, Negishi N, Hirano K, Yahata T, Ando K, Zuklys S, Holländer GA, Shima DT, Habu S (2008) Delta-like 4 is indispensable in thymic environment specific for T cell development. J Exp Med 205:2507–2513

Hu B, Lefort K, Qiu W, Nguyen BC, Rajaram RD, Castillo E, He F, Chen Y, Angel P, Brisken C, Dotto GP (2010) Control of hair follicle cell fate by underlying mesenchyme through a CSL-Wnt5a-FoxN1 regulatory axis. Genes Dev 24:1519–1532

Jerome LA, Papaioannou VE (2001) DiGeorge syndrome phenotype in mice mutant for the T-box gene. Tbx. Nat Genet 27:286–291

Klug DB, Carter C, Crouch E, Roop D, Conti CJ, Richie ER (1998) Interdependence of cortical thymic epithelial cell differentiation and T-lineage commitment. Proc Natl Acad Sci USA 95:11822–11827

Klug DB, Carter C, Gimenez-Conti IB, Richie ER (2002) Thymocyte-independent and thymocyte-dependent phases of epithelial patterning in the fetal thymus. J Immunol 169:2842–2845

Koch U, Fiorini E, Benedito R, Besseyrias V, Schuster-Gossler K, Pierres M, Manley NR, Duarte A, Macdonald HR, Radtke F (2008) Delta-like 4 is the essential, nonredundant ligand for Notch1 during thymic T cell lineage commitment. J Exp Med 205:2515–2523

Kondo M, Akashi K, Domen J, Sugamura K, Weissman IL (1997) Bcl-2 rescues T lymphopoiesis, but not B or NK cell development, in common gamma chain-deficient mice. Immunity 7:155–162

Kyewski BA, Kaplan HS (1982) Lymphoepithelial interactions in the mouse thymus: phenotypic and kinetic studies on thymic nurse cells. J Immunol 128:2287–2294

Laufer TM, DeKoning J, Markowitz JS, Lo D, Glimcher LH (1996) Unopposed positive selection and autoreactivity in mice expressing class II MHC only on thymic cortex. Nature 383:81–85

Lindsay EA, Vitelli F, Su H, Morishima M, Huynh T, Pramparo T, Jurecic V, Ogunrinu G, Sutherland HF, Scambler PJ, Bradley A, Baldini A (2001) Tbx1 haploinsufficiency in the DiGeorge syndrome region causes aortic arch defects in mice. Nature 410:97–101

Liu C, Saito F, Liu Z, Lei Y, Uehara S, Love P, Lipp M, Kondo S, Manley N, Takahama Y (2006) Coordination between CCR7- and CCR9-mediated chemokine signals in prevascular fetal thymus colonization. Blood 108:2531–2539

Maillard I, Tu L, Sambandam A, Yashiro-Ohtani Y, Millholland J, Keeshan K, Shestova O, Xu L, Bhandoola A, Pear WS (2006) The requirement for Notch signaling at the beta-selection checkpoint in vivo is absolute and independent of the pre-T cell receptor. J Exp Med 203:2239–2245

Manley NR, Capecchi MR (1995) The role of Hoxa-3 in mouse thymus and thyroid development. Development 121:1989–2003

Development and Function of Cortical Thymic Epithelial Cells

Maraskovsky E, O'Reilly LA, Teepe M, Corcoran LM, Peschon JJ, Strasser A (1997) Bcl-2 can rescue T lymphocyte development in interleukin-7 receptor-deficient mice but not in mutant rag-1$^{-/-}$ mice. Cell 89:1011–1019

Mat Ripen A, Nitta T, Murata S, Tanaka K, Takahama Y (2011) Ontogeny of thymic cortical epithelial cells expressing the thymoproteasome subunit β5t. Eur J Immunol 41:1278–1287

Mazzucchelli RI, Warming S, Lawrence SM, Ishii M, Abshari M, Washington AV, Feigenbaum L, Warner AC, Sims DJ, Li WQ, Hixon JA, Gray DH, Rich BE, Morrow M, Anver MR, Cherry J, Naf D, Sternberg LR, McVicar DW, Farr AG, Germain RN, Rogers K, Jenkins NA, Copeland NG, Durum SK (2009) Visualization and identification of IL-7 producing cells in reporter mice. PLoS One 4:7637

Merscher S, Funke B, Epstein JA, Heyer J, Puech A, Lu MM, Xavier RJ, Demay MB, Russell RG, Factor S, Tokooya K, Jore BS, Lopez M, Pandita RK, Lia M, Carrion D, Xu H, Schorle H, Kobler JB, Scambler P, Wynshaw-Boris A, Skoultchi AI, Morrow BE, Kucherlapati R (2001) TBX1 is responsible for cardiovascular defects in velo-cardio-facial/DiGeorge syndrome. Cell 104:619–629

Murata S, Sasaki K, Kishimoto T, Niwa S, Hayashi H, Takahama Y, Tanaka K (2007) Regulation of CD8$^+$ T cell development by thymus-specific proteasomes. Science 316:1349–1353

Murata S, Yashiroda H, Tanaka K (2009) Molecular mechanisms of proteasome assembly. Nat Rev Mol Cell Biol 10:104–115

Nakagawa T, Roth W, Wong P, Nelson A, Farr A, Deussing J, Villadangos JA, Ploegh H, Peters C, Rudensky AY (1998) Cathepsin L: critical role in Ii degradation and CD4 T cell selection in the thymus. Science 280:450–453

Nakagawa Y, Ohigashi I, Nitta T, Sakata M, Tanaka K, Murata S, Kanagawa O, Takahama Y (2012) Thymic nurse cells provide microenvironment for secondary TCRα rearrangement in cortical thymocytes. Proc Natl Acad Sci USA 109:20572–20577

Nedjic J, Aichinger M, Emmerich J, Mizushima N, Klein L (2008) Autophagy in thymic epithelium shapes the T-cell repertoire and is essential for tolerance. Nature 455:396–400

Neefjes J, Jongsma ML, Paul P, Bakke O (2011) Towards a systems understanding of MHC class I and MHC class II antigen presentation. Nat Rev Immunol 11:823–836

Nehls M, Pfeifer D, Schorpp M, Hedrich H, Boehm T (1994) New member of the winged-helix protein family disrupted in mouse and rat nude mutations. Nature 372:103–107

Nehls M, Kyewski B, Messerle M, Waldschutz R, Schüddekopf K, Smith AJH, Boehm T (1996) Two genetically separable steps in the differentiation of thymic epithelium. Science 272:886–889

Nishikawa Y, Hirota F, Yano M, Kitajima H, Miyazaki J, Kawamoto H, Mouri Y, Matsumoto M (2010) Biphasic Aire expression in early embryos and in medullary thymic epithelial cells before end-stage terminal differentiation. J Exp Med 207:963–971

Nitta T, Murata S, Sasaki K, Fujii H, Ripen AM, Ishimaru N, Koyasu S, Tanaka K, Takahama Y (2010) Thymoproteasome shapes immunocompetent repertoire of CD8$^+$ T cells. Immunity 32:29–40

Park JH, Adoro S, Guinter T, Erman B, Alag AS, Catalfamo M, Kimura MY, Cui Y, Lucas PJ, Gress RE, Kubo M, Hennighausen L, Feigenbaum L, Singer A (2010) Signaling by intrathymic cytokines, not T cell antigen receptors, specifies CD8 lineage choice and promotes the differentiation of cytotoxic-lineage T cells. Nat Immunol 11:257–264

Pellegrini M, Bouillet P, Robati M, Belz GT, Davey GM, Strasser A (2004) Loss of Bim increases T cell production and function in interleukin 7 receptor-deficient mice. J Exp Med 200:1189–1195

Peschon JJ, Morrissey PJ, Grabstein KH, Ramsdell FJ, Maraskovsky E, Gliniak BC, Park LS, Ziegler SF, Williams DE, Ware CB, Meyer JD, Davison BL (1994) Early lymphocyte expansion is severely impaired in interleukin 7 receptor-deficient mice. J Exp Med 180:1955–1960

Pezzano M, Samms M, Martinez M, Guyden J (2001) Questionable thymic nurse cell. Microbiol Mol Biol Rev 65:390–403

Radtke F, Wilson A, Stark G, Bauer M, van Meerwijk J, MacDonald HR, Aguet M (1999) Deficient T cell fate specification in mice with an induced inactivation of Notch1. Immunity 10:547–558

Radtke F, Wilson A, Mancini SJ, MacDonald HR (2004) Notch regulation of lymphocyte development and function. Nat Immunol 5:247–253

Rieker T, Penninger J, Romani N, Wick G (1995) Chicken thymic nurse cells: an overview. Dev Comp Immunol 19:281–289

Ritter MA, Sauvage CA, Cotmore SF (1981) The human thymus microenvironment: in vivo identification of thymic nurse cells and other antigenically-distinct subpopulations of epithelial cells. Immunology 44:439–446

Rock KL, Goldberg AL (1999) Degradation of cell proteins and the generation of MHC class I-presented peptides. Annu Rev Immunol 17:739–779

Rode I, Boehm T (2012) Regenerative capacity of adult cortical thymic epithelial cells. Proc Natl Acad Sci USA 109:3463–3468

Rossi SW, Jenkinson WE, Anderson G, Jenkinson EJ (2006) Clonal analysis reveals a common progenitor for thymic cortical and medullary epithelium. Nature 441:988–991

Rossi SW, Kim MY, Leibbrandt A, Parnell SM, Jenkinson WE, Glanville SH, McConnell FM, Scott HS, Penninger JM, Jenkinson EJ, Lane PJ, Anderson G (2007) RANK signals from CD4$^+$3$^-$ inducer cells regulate development of Aire-expressing epithelial cells in the thymic medulla. J Exp Med 204:1267–1272

Rudd BD, Venturi V, Li G, Samadder P, Ertelt JM, Way SS, Davenport MP, Nikolich-Žugich J (2011) Nonrandom attrition of the naive CD8+ T-cell pool with aging governed by T-cell receptor: pMHC interactions. Proc Natl Acad Sci USA 108:13694–13699

Schaller CE, Wang CL, Beck-Engeser G, Goss L, Scott HS, Anderson MS, Wabl M (2008) Expression of Aire and the early wave of apoptosis in spermatogenesis. J Immunol 180:1338–1343

Schmitt TM, Zúñiga-Pflücker JC (2002) Induction of T cell development from hematopoietic progenitor cells by delta-like-1 in vitro. Immunity 17:749–756

Shakib S, Desanti GE, Jenkinson WE, Parnell SM, Jenkinson EJ, Anderson G (2009) Checkpoints in the development of thymic cortical epithelial cells. J Immunol 182:130–137

Shinkura R, Kitada K, Matsuda F, Tashiro K, Ikuta K, Suzuki M, Kogishi K, Serikawa T, Honjo T (1999) Alymphoplasia is caused by a point mutation in the mouse gene encoding Nf-κb-inducing kinase. Nat Genet 22:74–77

Shortman K, Scollay R, Andrews P, Boyd R (1986) Development of T lymphocytes within the thymus and within thymic nurse cells. Curr Top Microbiol Immunol 126:5–18

Su D, Ellis S, Napier A, Lee K, Manley NR (2001) Hoxa3 and Pax1 regulate epithelial cell death and proliferation during thymus and parathyroid organogenesis. Dev Biol 236:316–329

Takahama Y, Letterio JJ, Suzuki H, Farr AG, Singer A (1994) Early progression of thymocytes along the CD4/CD8 developmental pathway is regulated by a subset of thymic epithelial cells expressing transforming growth factor β. J Exp Med 179:1495–1506

Takahama Y, Nitta T, Mat Ripen A, Nitta S, Murata S, Tanaka K (2010) Role of thymic cortex-specific self-peptides in positive selection of T cells. Sem Immunol 22:287–293

Takahama Y, Takada K, Murata S, Tanaka K (2012) β5t-containing thymoproteasome: specific expression in thymic cortical epithelial cells and role in positive selection of CD8$^+$ T cells. Curr Opin Immunol 24:92–98

Thompson PK, Zúñiga-Pflücker JC (2011) On becoming a T cell, a convergence of factors kick it up a Notch along the way. Semin Immunol 23:350–359

Tousaint-Demylle D, Scheiff JM, Haumount S (1990) Thymic nurse cells: morphological study during their isolation from murine thymus. Cell Tissue Res 261:115–123

van de Wijngaert FP, Rademakers LH, Schuurman HJ, de Weger RA, Kater L (1983) Identification and in situ localization of the "thymic nurse cell" in man. J Immunol 130:2348–2351

van Ewijk W, Shores EW, Singer A (1994) Crosstalk in the mouse thymus. Immunol Today 15:214–217

Viret C, Lamare C, Guiraud M, Fazilleau N, Bour A, Malissen B, Carrier A, Guerder S (2011a) Thymus-specific serine protease contributes to the diversification of the functional endogenous CD4 T cell receptor repertoire. J Exp Med 208:3–11

Viret C, Leung-Theung-Long S, Serre L, Lamare C, Vignali DA, Malissen B, Carrier A, Guerder S (2011b) Thymus-specific serine protease controls autoreactive CD4 T cell development and autoimmune diabetes in mice. J Clin Invest 121:1810–1821

von Freeden-Jeffry U, Solvason N, Howard M, Murray R (1997) The earliest T lineage-committed cells depend on IL-7 for Bcl-2 expression and normal cell cycle progression. Immunity 7:147–154

Wada H, Masuda K, Satoh R, Kakugawa K, Ikawa T, Katsura Y, Kawamoto H (2008) Adult T-cell progenitors retain myeloid potential. Nature 452:768–772

Weiner L, Han R, Scicchitano BM, Li J, Hasegawa K, Grossi M, Lee D, Brissette JL (2007) Dedicated epithelial recipient cells determine pigmentation patterns. Cell 130:932–942

Wekerle H, Ketelson UP (1980) Thymic nurse cells. Ia-bearing epithelium involved in T-lymphocyte differentiation? Nature 283:402–404

Wekerle H, Ketelson UP, Ernst M (1980) Thymic nurse cells. lymphoepithelial cell complexes in murine thymuses: morphological and serological characterization. J Exp Med 151:925–944

Wick G, Rieker T, Penninger J (1991) Thymic nurse cells: a site for positive selection and differentiation of T cells. Curr Top Microbiol Immunol 173:99–105

Wilson A, MacDonald HR, Radtke F (2001) Notch 1-deficient common lymphoid precursors adopt a B cell fate in the thymus. J Exp Med 194:1003–1012

Wolfer A, Wilson A, Nemir M, MacDonald HR, Radtke F (2002) Inactivation of Notch1 impairs VDJ beta rearrangement and allows pre-TCR-independent survival of early alpha beta lineage thymocytes. Immunity 16:869–879

Yamasaki S, Saito T (2007) Molecular basis for pre-TCR-mediated autonomous signaling. Trends Immunol 28:39–43

Yang SJ, Ahn S, Park CS, Holmes KL, Westrup J, Chang CH, Kim MG (2006) The quantitative assessment of MHC II on thymic epithelium: implications in cortical thymocyte development. Int Immunol 18:729–739

Zamisch M, Moore-Scott B, Su D, Lucas PJ, Manley N, Richie ER (2005) Ontogeny and regulation of IL-7-expressing thymic epithelial cells. J Immunol 174:60–67

Zou D, Silvius D, Davenport J, Grifone R, Maire P, Xu PX (2006) Patterning of the third pharyngeal pouch into thymus/parathyroid by Six and Eya1. Dev Biol 293:499–512

Mechanisms of Thymus Medulla Development and Function

Graham Anderson, Song Baik, Jennifer E. Cowan,
Amanda M. Holland, Nicholas I. McCarthy, Kyoko Nakamura,
Sonia M. Parnell, Andrea J. White, Peter J. L. Lane,
Eric J. Jenkinson and William E. Jenkinson

Abstract The development of CD4$^+$ helper and CD8$^+$ cytotoxic T-cells expressing the $\alpha\beta$ form of the T-cell receptor ($\alpha\beta$TCR) takes place in the thymus, a primary lymphoid organ containing distinct cortical and medullary microenvironments. While the cortex represents a site of early T-cell precursor development, and the positive selection of CD4$^+$8$^+$ thymocytes, the thymic medulla plays a key role in tolerance induction, ensuring that thymic emigrants are purged of autoreactive $\alpha\beta$TCR specificities. In recent years, advances have been made in understanding the development and function of thymic medullary epithelial cells, most notably the subset defined by expression of the Autoimmune Regulator (Aire) gene. Here, we summarize current knowledge of the developmental mechanisms regulating thymus medulla development, and examine the role of the thymus medulla in recessive (negative selection) and dominant (T-regulatory cell) tolerance.

Contents

1	Introduction	20
2	Cellular Features of the Thymic Medulla	21
	2.1 Medullary Hemopoietic Non-T Lineage Cells	21
	2.2 Medullary Thymic Epithelial Cells	22
	2.3 Non-epithelial Mesenchymal Stroma	24
3	Development of Thymic Medullary Epithelium	25
	3.1 Defining mTEC Progenitors	25
	3.2 Cellular and Molecular Regulation of the mTEC Compartment	26

G. Anderson (✉) · S. Baik · J. E. Cowan · Amanda M. Holland · N. I. McCarthy ·
K. Nakamura · SoniaM. Parnell · A. J. White · P. J. L. Lane ·
EricJ. Jenkinson · W. E. Jenkinson
MRC Centre for Immune Regulation, Institute for Biomedical Research, Medical School,
University of Birmingham, Edgbaston, Birmingham B15 2TT, UK
e-mail: g.anderson@bham.ac.uk

Current Topics in Microbiology and Immunology (2014) 373: 19–47
DOI: 10.1007/82_2013_320
© Springer-Verlag Berlin Heidelberg 2013
Published Online: 24 April 2013

4	Functions of the Thymic Medulla	29
	4.1 Medullary Thymic Epithelium and Central Tolerance Induction	29
	4.2 Foxp3 Regulatory T-Cell Development	33
	4.3 Post-Selection Thymocyte Differentiation	36
5	Conclusions	39
References		39

1 Introduction

T-cells bearing the alpha–beta T-cell receptor complex ($\alpha\beta$TCR) represent a critical cellular component of immune responses aimed at targeting a wide range of pathogens including bacteria and viruses. The development of $\alpha\beta$T-cells occurs within the thymus, a process that is initiated following the entry of blood-borne lymphoid progenitors originating from the fetal liver or bone marrow (Anderson et al. 2007; Takahama 2006). Intrathymic T-cell development is a complex process, and involves a series of steps including T-cell commitment, proliferation, differentiation, selection, and migration. To accommodate this developmental program, the thymus consists of distinct T-cellular microenvironments in which thymocytes at particular developmental stages are housed. For example, immature T-cell precursors defined by their lack of expression of CD4 and CD8 are enriched in the subcapsular region, while their CD4$^+$8$^+$ progeny, representing the bulk of thymocytes, reside within the thymic cortex. In contrast, the thymus medulla provides a microenvironment for the most mature single positive (SP) CD4$^+$ and CD8$^+$ cells expressing high levels of the $\alpha\beta$TCR. Importantly, these major regions of the thymus are further defined by the phenotypically and functionally distinct stromal cells that are contained within them, including cortical thymic epithelium (cTEC) and medullary thymic epithelium (mTEC) (Alves et al. 2009).

Current models of thymic function are based upon the idea that an ordered process of T-cell development occurs as a result of the sequential migration of developing thymocytes through these distinct stromal microenvironments, ensuring that they receive important signals and cell–cell interactions in an appropriate order and context (Petrie and Zuniga-Pflucker 2007). The primary aim of this review is to discuss the role of the thymus medulla in $\alpha\beta$T-cell development. In particular, we will summarize the current knowledge of the cellular and molecular interactions that lead to thymic medulla formation, focusing on the processes involving maturation of mTEC. In addition, we will examine how thymic medullary environments contribute to both deletional and dominant self-tolerance mechanisms that operate upon the newly positively selected $\alpha\beta$TCR repertoire.

2 Cellular Features of the Thymic Medulla

2.1 Medullary Hemopoietic Non-T Lineage Cells

While the major hemopoietic compartment of the thymic medulla consists of $CD4^+$ and $CD8^+$ $\alpha\beta TCR^{hi}$ thymocytes generated as a result of positive selection in the thymic cortex, it also contains a variety of hemopoietic accessory cells that are linked to its function. Notably, thymic dendritic cells (tDC) are enriched in medullary areas and at the surrounding cortico-medullary junction. Given the role of tDC in purging the positively selected repertoire of potentially autoreactive specificities, such positioning is likely to be of importance in the screening of newly selected cells as they migrate from the cortex into the medulla. Interestingly, tDC are heterogeneous, suggestive of differing roles in thymocyte differentiation. Thus, in the adult thymus, three phenotypically distinct tDC subsets have been identified, namely plasmacytoid DC (pDC), and two subsets of conventional DC (cDC) that can be defined by $CD8\alpha^{low}CD11b^+SIRP\alpha^+$ and $CD8\alpha^{highCD11b-}SIRP\alpha^-$ phenotypes. Interestingly, these distinct tDC subsets have distinct developmental origins—while $SIRP\alpha^-$ tDC are generated intrathymically from immature progenitors, both $SIRP\alpha^+$ tDC and pDC are recruited to the thymus from the periphery. Despite the known heterogeneity of tDC in the thymus, relatively little is known about their anatomical location and positioning, and the long-held view is that their location is limited to the medulla and surrounding cortico-medullary junction. Interestingly however, a study recently showed that despite the presence of abundant medullary-resident $CD11c^+$ tDC, $SIRP\alpha^+$ tDC were notably absent from the medulla, and instead could be detected within thymic cortical regions, often in association with small vessels (Baba et al. 2009). Indeed, two-photon microscopy of explanted thymic tissue demonstrated the formation of interactions between thymocytes and tDC within the thymic cortex, again at regions containing capillaries (Ladi et al. 2008). Collectively, such observations argue against the notion that tDC are restricted to medullary regions and instead suggest that distinct tDC subsets can be specifically positioned within particular regions of the thymus, including the cortex. Moreover, multiple chemokine receptors have been highlighted in relation to tDC recruitment and positioning, including CCR2 (Baba et al. 2009), CCR7 (Ladi et al. 2008), CCR9 (Hadeiba et al. 2012), and XCR1 (Lei et al. 2011), suggesting that chemokine production from distinct intrathymic microenvironments is important in the context of tDC location and function.

While tDC are important mediators of intrathymic negative selection of autoreactive thymocytes, other hemopoietic accessory cells are directly linked to the development of thymic medullary microenvironments. In particular, Lymphoid Tissue inducer (LTi) cells are present within thymic medullary regions, and through their provision of TNFSF ligands such as RANKL, have been shown to stimulate the maturation of $RANK^+$ mTEC progenitors (Rossi et al. 2007). Perhaps importantly, LTi cells, first reported as essential mediators of lymph node (LN) organogenesis in the embryonic period (Cupedo et al. 2002), are found in both the fetal and adult thymus in close association with mTEC. Moreover, analysis of LTi-deficient ROR$\gamma^{-/-}$ mice at embryonic stages prior to the emergence of positively selected $\alpha\beta$TCR^{hi} thymocytes has provided direct evidence that LTi cells are key to the generation of the first cohorts of Aire$^+$ mTEC (White et al. 2008), the development of which represents a critical step in the establishment of T-cell tolerance in the neonatal period (Guerau-de-Arellano et al. 2009). Unlike their well-documented role in fetal thymus, ascribing a specific role to LTi in the adult thymus has been difficult, particularly since mTEC abnormalities in RORγ-deficient mice could also be explained by defective $\alpha\beta$T-cell development. However, a recent study (Dudakov et al. 2012) showed a link between LTi and regeneration of the adult thymus following experimental ablation. Thus, irradiation-induced thymic atrophy resulted in the enhanced production of IL-22 by RORγt$^+$CC R6$^+$NKp46$^-$ LTi cells, with IL-22 then operating directly on thymic epithelial compartments to promote their expansion. Given that the mTEC lineage can be separated into distinct developmental stages (Dooley et al. 2008; Gabler et al. 2007; Nishikawa et al. 2010; Rossi et al. 2007), while stages in the cTEC lineage have also been described (Nowell et al. 2011; Ripen et al. 2011; Shakib et al. 2009), it will be interesting to determine whether IL-22 exerts its effect on immature or mature TEC populations, or both. Finally, although thymic LTi have been shown to have shared a common RORγt$^+$CD4$^+$IL7Rα^+RANKL$^+$ phenotype with LTi in peripheral lymphoid tissues (Anderson et al. 2007), it is currently unclear whether thymus and LN harbor tissue-specific LTi subsets, or whether LTi populations are capable of trafficking between these tissues.

2.2 Medullary Thymic Epithelial Cells

Immunohistological analysis of thymic microenvironments is a widely used approach with which to dissect the cellular complexity of cortical and medullary areas, enabling the phenotypic definition of stromal compartments in both areas, most notably thymic epithelial cells (TEC). Tissue sections of adult thymus often show individual medullary regions embedded within a cortical matrix, although it is important to note that the thymus medulla as a whole represents a complex structure with seemingly separate medullary areas actually joined by interconnecting branches (Anderson et al. 2000). While individual medullary areas have been shown to occur as a result of the expansion and differentiation of single mTEC progenitors (Rodewald et al. 2001), it is not clear how the complex three-dimensional organization of the thymic medulla is controlled, although thymic vasculature has been proposed to play a role (Anderson et al. 2000).

The cTEC and mTEC compartments are identified by both shared and lineage restricted molecules (Fig. 1). Many of the reagents that are used to define TEC immunohistologically, in addition to the pan-epithelial marker EpCAM1 (Nelson et al. 1996), recognize cytokeratin family members, structural proteins that reflect

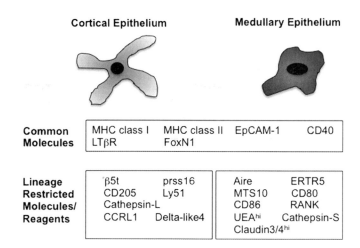

Fig. 1 Shared and lineage restricted markers of cortical and medullary thymic epithelial cells. Panels of markers used in both immunohistochemical and flow cytometric analysis are frequently used to study cTEC and mTEC lineages. While some molecules are common to both, others enable the discrimination of these discrete lineages. However, it is important to note that it is currently unclear how expression of these markers relates to distinct immature progenitors and mature stages within TEC lineages

the differing morphology of cTEC and mTEC compartments (Farr and Braddy 1989). Thus, unlike their cTEC counterparts, mTEC are often defined by expression of cytokeratin-5 and cytokeratin-14, and lack of expression of cytokeratin-8/18 (Klug et al. 1998, 2002). In addition, antibodies that recognize unknown molecules expressed by mTEC include ERTR5 (Van Vliet et al. 1984) and MTS10 (Godfrey et al. 1990), while the fucose binding lectins Tetragonolobus Purpureas Agglutinin (TPA), and Ulex Europeus Agglutinin (UEA) also demonstrate selective reactivity with the thymic medulla in tissue sections (Farr and Anderson 1985). However, it is not entirely clear from this type of tissue section analysis whether such reagents reflect 'pan-mTEC' markers that react with the whole mTEC compartment, or whether distinct mTEC subsets exist. Perhaps importantly, immunohistochemical analysis of the mTEC compartment can be further complemented by flow cytometric analysis of enzymatically disaggregated thymus preparations. Although analysis of TEC compartments following enzymatic digestion can be limited by the sensitivity of cell surface molecules (Izon et al. 1994; Seach et al. 2012), a panel of markers has emerged that is now widely used in association with enzymatic digestion. Thus, total TEC are frequently defined as $CD45^-EpCAM1^+$, which can be further subdivided on the basis of cell surface expression of Ly51, enabling the discrimination of $Ly51^+$ cTEC and $Ly51^-$ mTEC compartments. Within the mTEC lineage, an additional panel of molecules including CD40, CD80, MHC class II, and Aire reveal distinct subsets including $CD80^-MHCII^{low}$ and $CD80^+MHCII^{hi}$ cells, often referred to as $mTEC^{low}$ and $mTEC^{hi}$ (Derbinski et al. 2001; Gray et al. 2006; Hubert et al. 2008;

Rossi et al. 2007). Until recently, the relevance of phenotypically distinct mTEC subsets was not clear. However, many studies now show that mTEC represent a dynamic thymic compartment, that can be defined by precursor-product relationships with a turnover time of 2–3 weeks for the mature mTEC population (Gabler et al. 2007). Functional analysis of the developmental relationships of distinct mTEC subsets will be discussed in Sect. 3.

2.3 Non-epithelial Mesenchymal Stroma

The thymus is an epithelial–mesenchymal tissue, and during early stages of thymus organogenesis, the inner epithelial rudiment is surrounded by a layer of mesenchyme derived from the neural crest (Manley and Blackburn 2003; Rodewald 2008). Within the adult thymus, several cell fate-mapping studies have now shown that much of the mesenchyme present is of neural crest origin, where it is associated with epithelium and the endothelium of the thymus vasculature (Foster et al. 2008; Muller et al. 2008; Yamazaki et al. 2005). During thymus development, mesenchymal cells that form the thymic capsule penetrate the epithelial core, separating it into lobules via trabeculae. In addition to its mesenchymal components, vascularization of the developing thymus occurs after anlage formation, culminating in a complex network of both blood and lymphatic vessels (Odaka et al. 2006) that are composed of perivascular cells and endothelium. Thus, a panel of markers including smooth muscle actin, ERTR7, and desmin has been used to define histological organization of non-epithelial medulla stroma (Odaka 2009), while flow cytometric analysis using the markers podoplanin, Ly51, and PDGFRα reveals complex heterogeneity in mesenchymal subsets (Jenkinson et al. 2007; Muller et al. 2005). While further analysis of the functional importance of these distinct compartments is required, it is interesting to note that thymic mesenchyme can act as both positive and negative regulators of TEC expansion, through their control of the Retinoic Acid and Fibroblast Growth Factor pathways (Jenkinson et al. 2003; Sitnik et al. 2012).

The corticomedullary junction (CMJ) represents an important area with respect to vasculature, with both the entry of lymphoid progenitors and the exit of mature thymocytes taking place at this site (Porritt et al. 2003). Indeed, the perivascular spaces of blood vessels at the CMJ contain c-Kit[+] T-cell precursors and CD4[+] and CD8[+] thymocytes (Mori et al. 2007), with neural crest derived pericytes controlling the emigration of the latter via their production of sphingosine-1-phosphate (S1P), a ligand for sphingosine-1-phosphate receptor-1 (S1PR1) expressed by mature thymocytes (Zachariah and Cyster 2010). Additionally, a non-epithelial conduit system has been identified in human thymus, which represents a network of inter-connecting tubules containing multiple basement membrane components including laminin-5, collagen type IV and perlecan (Drumea-Mirancea et al. 2006). Interestingly, such a network is reminiscent of the conduit system present within the T-zone of the lymph node and spleen, further highlighting the similarities between the thymic medulla and compartments within secondary lymphoid tissues (Derbinski and Kyewski 2005). While the functional importance of the thymic medullary conduit system remains unclear, its diameter appears too small to enable transport of cells (Drumea-Mirancea et al. 2006), leaving open the possibility that by acting as a transport network for small molecules such as antigen and chemokines, it plays a role in medullary thymocyte migration and tolerance induction.

3 Development of Thymic Medullary Epithelium

3.1 Defining mTEC Progenitors

Although the mTEC compartment has been shown to share a common bipotent progenitor with the cTEC (Bleul et al. 2006; Rossi et al. 2006), relatively little is known about the mechanisms controlling the emergence of cells that are committed to the mTEC lineage from this progenitor pool. Recently however the possible role of FoxN1, a transcription factor representing a master regulator of TEC differentiation (Blackburn et al. 1996; Nehls et al. 1994, 1996), has been evaluated through analysis of TEC development in FoxN1-deficient nude mice and a panel of mice expressing FoxN1 at varying levels (Nowell et al. 2011). Interestingly, these findings suggested that the mTEC lineage might emerge from the bipotent TEC progenitor stage via a mechanism occurring independently of FoxN1. Given that bipotent TEC progenitors persist within the FoxN1-deficient thymus rudiment at least until the postnatal stages (Bleul et al. 2006), these findings suggest that FoxN1 may be selectively required downstream of the emergence of mTEC progenitors, perhaps through controlling their survival as well as differentiation.

The first data demonstrating the existence of mTEC committed progenitors involved functional clonal analyses in the absence of a defined phenotype (Rodewald et al. 2001). Subsequent attempts to define and then directly isolate mTEC committed progenitors have often relied upon use of markers typically associated with the mature mTEC lineage in the context of the developing embryonic thymus, so the accurate phenotype of these cells, and the separation of immature and mature mTEC remains obscure. For example, claudin-3 and claudin-4, tight junction components expressed by mTEC in the adult thymus, have been shown to identify TEC within the early thymus anlage that are also reactive with the mTEC markers MTS10 and UEA1 (Hamazaki et al. 2007). Perhaps most importantly, purified Claudin3/4^{hi} embryonic TEC were shown to give rise to mature Aire$^+$ mTEC in precursor-product experiments involving reaggregate thymus organ cultures (RTOC), providing the first phenotypic definition of mTEC progenitors (Hamazaki et al. 2007). In other studies, analysis of the mTEC compartment using CD80 and MHCII expression showed that during embryonic thymus development, CD80$^-$MHCII^{low} 'mTEC^{low}' cells appear prior to the emergence of CD80$^+$MHCII^{hi} 'mTEC^{hi}' cells, suggesting a possible precursor-product relationship between these populations (Gabler et al. 2007; Rossi et al. 2007). Importantly, direct analysis of mTEC development using RTOC experiments demonstrated that mTEC^{low} were able to give rise to their more mature mTEC^{hi} counterparts, including the subset expressing Aire (Gabler et al. 2007; Rossi et al. 2007). Further, mTEC^{low} and mTEC^{hi} subsets are also present in the adult thymus (Gray et al. 2006), with BrdU labeling experiments providing evidence of the continued generation of mTEC^{hi} from mTEC^{lo} cells in the postnatal thymus, with a turnover time of 2–3 weeks for mTEC^{hi} cells (Gabler et al. 2007). Collectively, these studies were important as they highlighted distinct developmental stages within mTEC, and provided direct indications that the epithelial component of the thymus medulla represents a dynamic cellular microenvironment undergoing constant renewal. Importantly however, it is perhaps important to note that precursor-product analysis of mTEC has frequently focused on events that culminate in generation of the Aire$^+$ subset. Thus, it remains possible that other mature mTEC subsets exist that are not linked to the same Aire-expressing pathway, and which could be generated via a separate mTEC progenitor pool. Whether such a subset resides within the mTEC^{low} population requires a more detailed phenotypic and functional analysis of these poorly defined cells.

3.2 Cellular and Molecular Regulation of the mTEC Compartment

A normal program of T-cell development and selection depends upon sequential interactions between thymocytes and stromal cells in the cortex and then the medulla. Importantly, studies in the late 1980s provided the first indications that growth and formation of the thymic medulla was, in turn, influenced by developing thymocytes. For example, analysis of thymic microenvironments following disruption of thymic hemopoietic compartments by either irradiation (Adkins et al. 1988) or treatment with the immunosuppressant cyclosporin A (Kanariou et al. 1989) was shown to have a dramatic, and reversible, impact on mTEC. Critically, subsequent experiments showed that the transplantation of WT hemopoietic progenitors into SCID mice corrected their severely disorganized thymic epithelial microenvironments (Shores et al. 1991), providing the first evidence that signals from hemopoietic cells directly influenced thymic epithelial cell development. Other studies showed that peripheral T-cells (Surh et al. 1992) and SP thymocytes could also regulate the mTEC compartment, a process requiring $\alpha\beta$TCR expression (Palmer et al. 1993; Shores et al. 1994). Such observations were collectively described as a 'thymus crosstalk' process, (van Ewijk et al. 1994), during which interaction with, and signals from, developing thymocytes are required for the formation of thymic epithelial microenvironments.

Although the studies above provided information on the cellular source of the molecules that promote mTEC development and medullary growth, the nature of the signals provided by developing thymocytes and/or additional hemopoietic cells was, until recently poorly understood. However, several studies had noted that mice harboring mutations in several genes critical in the NF-κB signaling, including TRAF6 (Akiyama et al. 2005), NIK (Kajiura et al. 2004), and RelB (Burkly et al. 1995; Heino et al. 2000; Weih et al. 1995; Zuklys et al. 2000) displayed mTEC abnormalities. Such phenotypes often included reduced/absent Aire expression and a failure to establish T-cell tolerance, suggesting that cell surface receptors expressed by mTEC that utilize the NF-kB signaling cascade could be critical molecular components of thymus medulla crosstalk. Interestingly, the development of secondary lymphoid tissues is known to involve multiple members of the Tumor Necrosis Factor Receptor SuperFamily (TNFRSF) (Weih and Caamano 2003), whose ligands are expressed by lymphoid cells, raising the possibility that a similar axis might also be involved in formation of medullary thymic microenvironments (Anderson et al. 2007; Derbinski and Kyewski 2005). Indeed, many studies have now shown the expression of various TNFRSF members by mTEC, some of which have been shown to play a direct role during mTEC development. Of these, Lymphotoxinβ Receptor (LTβR, TNFRSF3), CD40 (TNFRSF5), and RANK (TNFRSF11a) remain the best studied. For example, studies on LTβR$^{-/-}$ mice have demonstrated a reduction in mTEC numbers and medullary disorganization, which is associated with abnormalities in thymocyte emigration and autoimmunity. Importantly, although initial studies (Chin et al. 2003) suggested that the LT-LTβR axis was linked to the generation of Aire-expressing mTEC, other studies showed this not to be the case (Martins et al. 2008; Venanzi et al. 2007). Rather, LTβR appears to be involved in mechanisms controlling the expression of Aire-independent Tissue Restricted Antigens (TRAs), as well as the chemokines CCL19 and CCL21 (Chin et al. 2006; Seach et al. 2008; Zhu et al. 2007). Importantly however, as well as being expressed by mTEC, LTβR is also detectable within cTEC and MTS15$^+$ fibroblasts (Hikosaka et al. 2008; Seach et al. 2008). So, it remains unclear which features of LTβR deficiency are a direct result of absence of LTβR expression by mTEC, or whether abnormalities can occur indirectly as a result of absence of expression in other thymic stromal compartments. In relation to the involvement of LTβR in thymocyte-TEC crosstalk, several studies now show that LTα and LTβ are expressed by mature SP thymocytes as compared to their CD4$^+$8$^+$ precursors (Boehm et al. 2003; White et al. 2008), indicating a crosstalk process involving positively selected thymocytes. Interestingly however, the absence of LIGHT, an additional LTβR ligand does not appear to play a role in mTEC development. Moreover, mTEC abnormalities in LTβR$^{-/-}$ mice are more severe as compared to LT$\beta^{-/-}$LIGHT$^{-/-}$ double deficient mice (Boehm et al. 2003), perhaps suggesting additional unknown ligands for LTβR that are expressed by thymocytes and which operate during mTEC development.

Both CD40 and RANK have been shown to play key roles in the generation of the Aire$^+$ subset of mTEC. While CD40 is expressed by both cTEC and mTEC,

RANK expression is higher in the latter (Hikosaka et al. 2008; Rossi et al. 2007; Shakib et al. 2009). Moreover, absence of RANK expression leads to a dramatic reduction in the frequency of Aire$^+$ mTEC, in both the fetal and adult thymus, with combined RANK/CD40 deficiency in the adult reducing this mTEC compartment further (Akiyama et al. 2008; Rossi et al. 2007). RANK deficiency and RANK/CD40 double deficiency also results in the onset of T-cell mediated autoimmunity, highlighting the importance of these TNFRSF molecules during intrathymic tolerance induction (Akiyama et al. 2008; Rossi et al. 2007). Several studies have investigated the cellular sources of RANKL and CD40L in relation to thymocyte crosstalk and thymic medulla formation, in both the fetal and adult thymus (Anderson and Takahama 2012). In the fetal thymus, we showed that RANKL expression maps to subsets of cells belonging to the innate immune system, including RORγt$^+$ Lymphoid Tissue Inducer (LTI) cells (Rossi et al. 2007), and invariant Vγ5$^+$TCR dendritic epidermal T-cell progenitors (Roberts et al. 2012). Interestingly, the involvement of the innate immune system during thymus medulla formation is active at developmental stages prior to positive selection of the $\alpha\beta$TCR repertoire (White et al. 2008), meaning that the crosstalk processes in the fetal and adult thymus medulla are distinct. Given the importance of Aire expression during neonatal tolerance (Guerau-de-Arellano et al. 2009), these findings suggest a scenario in which the innate immune system helps in the control of tolerance induction of the nascent TCR repertoire by ensuring efficient generation of mTEC compartments in the embryo.

In the adult thymus, as with LTβR ligands, positively selected thymocytes in particular CD4$^+$8$^-$ cells, act as sources of RANKL and CD40L (Hikosaka et al. 2008; Irla et al. 2008). Relevant to this, we have recently shown (Desanti et al. 2012) that RANKL and CD40L expression map to different subsets and developmental stages within the intrathymic CD4$^+$8$^-$ compartment. Thus, recently selected CD69$^+$CD4$^+$8$^-$ cells are enriched for RANKL$^+$ cells, while CD40L expression is linked to more mature CD69$^-$CD4$^+$8$^-$ thymocytes. Moreover, FoxP3$^+$ Regulatory T-cells present in the thymus express RANKL but not CD40L, demonstrating that thymocyte crosstalk in the development of the mTEC compartment involves distinct interactions with multiple CD4$^+$8$^-$ subsets.

While the above studies highlight cellular and molecular control involving the generation of Aire$^+$ mTEC, and although the CD80$^+$ mTEC subset which contains Aire expressing cells (Gray et al. 2006), have a turnover time of 2–3 weeks (Gabler et al. 2007), less is known about events occurring during late stage mTEC development. Indeed, uncertainty exists regarding possible stages of mTEC development post-Aire expression, and the role of Aire itself during mTEC development. Several models have been put forward to explain mTEC developmental programmes in relation to the timing of Aire expression. For example, based on cell fate mapping studies utilizing Aire-Cre transgenic mice, Aire$^+$CD80$^+$ mTEC were shown to progress to an Aire$^-$CD80^{low} stage (Nishikawa et al. 2010), suggestive of mTEC maturation post-Aire expression, and arguing against the idea that Aire directly promotes mTEC apoptosis (Gray et al. 2007). Interestingly, other studies described a small subset of Aire$^-$ mTEC that expressed involucrin, a marker of both keratinocyte terminal differentiation and Hassalls Corpuscles, concentric whorls of cells thought to represent end-stage medullary epithelium (Yano et al. 2008). Given that the frequency of involucrin$^+$ mTEC are reduced in the Aire$^{-/-}$ thymus (Yano et al. 2008), such observations support the 'Terminal Differentiation' model of mTEC development, in which Aire plays a role during end stage mTEC development (Matsumoto 2011). Indeed, involucrin$^+$ mTEC were reduced in the absence of LTβR signaling (White et al. 2010), suggesting further crosstalk mechanisms operating during post-Aire expression in mTEC. However, it is also important to note that other studies analyzing the disruption of mTEC development in Aire$^{-/-}$ mice have suggested that Aire is required during earlier stages of the mTEC developmental program (Dooley et al. 2008). In this 'Developmental Model', Aire may be involved in regulating the developmental choice of mTEC progenitors, a process that then impacts upon TRA expression within the thymic medulla (Gillard and Farr 2005). For example, Aire controls mTEC expression of a panel of transcription factors, including Oct4 and Nanog, typically associated with progenitor cells (Gillard et al. 2007). Importantly, while these models provide distinct views on the timing of Aire expression in the mTEC lineage, they collectively highlight the importance of Aire during normal thymus medulla development. While further studies are required, recent analyses suggest that Aire expression by mTEC is limited to a single window of 1–2 days (Wang et al. 2012), which may help to provide a clearer picture of the timing and role of Aire expression in relation to early and late mTEC developmental stages.

4 Functions of the Thymic Medulla

4.1 Medullary Thymic Epithelium and Central Tolerance Induction

TCR gene recombination occurs in a seemingly random manner leading to the generation of a highly diverse T-cell repertoire. While this provides a clear benefit in terms of the capacity of T-cells to recognize and respond to diverse pathogenic challenge, a potential negative aspect of this mode of TCR determination lies in the generation of T-cells bearing receptors capable of both recognizing and becoming activated by self-antigen. T-cell activation against antigens expressed by tissues of the host leads to the highly undesirable outcome of T-cell orchestrated autoimmune disease. In order to combat the potentially destructive generation and subsequent escape of autoreactive T-cells into the periphery, thymic medullary microenvironments provide several layers of tolerance induction, including that of deletional tolerance, acting to purge autoreactive T-cell clones from the naïve repertoire via negative selection (Fig. 2).

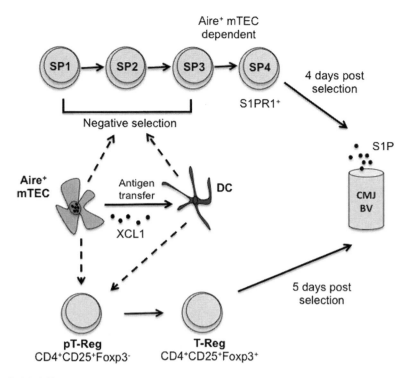

Fig. 2 Medullary thymic microenvironments regulate CD4 thymocyte maturation and selection via multiple mechanisms. Newly selected CD4$^+$8$^-$ T-cells interact with both Aire$^+$ mTEC and thymic dendritic cells during medullary maturation. Conventional CD4 thymocytes undergo a series of maturational steps (SP1-4), where semi-mature autoreactive T-cell clones are deleted at an immature stage (SP1-3) in part via Aire-dependent tissue restricted antigens generated and presented either directly by mTEC or indirectly via antigen transfer and presentation by thymic DC. Generation of SP4 CD4 thymocytes relies upon Aire$^+$ mTEC and subsequent emigration occurs 4 days post-selection in an S1P-dependent manner via blood vessels at the CMJ BV. Medullary APC interaction additionally drives pT-Reg induction in a TCR- and CD80/86-dependent manner leading to the generation of Foxp3$^+$ T-Reg that exit thymus 5 days post-selection

As described previously, the key cellular mediators of intrathymic central tolerance induction primarily constitute medullary thymic epithelium and thymic dendritic cells acting in concert. The clear requirement for medullary thymic epithelium in the induction of central tolerance via negative selection of autoreactive thymocytes is evidenced from several mutant mouse models demonstrating defective mTEC development and associated onset of autoimmune disease (Akiyama et al. 2005; Burkly et al. 1995; Nitta et al. 2011; Rossi et al. 2007). Following positive selection, maturing thymocytes demonstrate directed migration into medullary microenvironments where they spend a calculated 4–5 days scanning both mTEC and dendritic cells (McCaughtry et al. 2007). Interestingly, self-antigen presented to developing thymocytes includes multiple sources, comprising both peripheral self-antigen transported into the thymus by peripheral DC subsets and self-Ag generated from intrathymic microenvironments (Baba et al. 2009; Hadeiba et al. 2012; Klein et al. 2011).

Pivotal to the efficient role of mTEC in screening developing T-cells for autoreactive specificities is the precise array of self-antigens expressed by mTEC against which T-cells are tested for high affinity recognition and subsequent deletion. A key paradox in the intrathymic screening of TCR specificities is how T-cells, while anatomically restricted to thymic microenvironments during development, are exposed to the breadth of self-antigens normally associated with particular peripheral tissues. A series of refined experiments have gradually begun to unravel this contradiction of anatomical compartmentalization of T-cells and peripheral self-antigens. Primarily, it was discovered that mTEC possessed a highly unusual characteristic of being able to express a diverse array of antigens normally associated with defined peripheral tissues (Derbinski et al. 2001). This remarkable ability of mTEC to mimic the antigen profile of an array of different tissue types led to the search for specific molecular mechanisms regulating this unusual functional capacity. Significantly, several lines of evidence led to the discovery of the role of the transcriptional regulator Aire (Auto-Immune REgulator) in the control of mTEC peripheral tissue antigen expression. Notably, human patients demonstrating a mutation in Aire exhibit autoimmune disease, termed autoimmune polyglandular syndrome type-1 (APS-1) or autoimmune polyendocrinopathy-candidiasis-ectodermal dystrophy (APECED) (Mathis and Benoist 2009). Generation of mutant mouse strains lacking fully functional Aire protein demonstrated a broadly similar spectrum of autoimmune disease manifestation, providing a useful murine model to study the role of Aire in the appearance of associated autoimmune disease (Anderson et al. 2002; Ramsey et al. 2002). Investigation of the cellular expression pattern of Aire showed it was primarily restricted to thymic tissue, and moreover intrathymic expression was limited to a sub-population of mTEC (Heino et al. 2000; Nagamine et al. 1997). Importantly, lack of Aire expression by mTEC in murine knockout models directly resulted in reduced expression of specific peripheral tissue antigens and resulted in the generation of targeted autoimmune disease (Anderson et al. 2002; Ramsey et al. 2002). Direct evidence that mTEC Aire-mediated expression of ectopic peripheral tissue antigens played a central role in thymic deletional tolerance was subsequently demonstrated whereby deletion of potentially autoreactive transgenic T-cell clones capable of recognizing pancreatic-associated self-antigen with a high degree of affinity was driven in an Aire-dependent manner (Liston et al. 2003). The exquisite sensitivity of mTEC-mediated deletion of autoreactive T-cells was later demonstrated via elegant experiments demonstrating that the reduced expression of a single restricted antigen specifically associated with ocular tissue by mTEC could manifest in highly targeted auto-immunity targeted toward the eye (DeVoss et al. 2006).

It would therefore appear clear that mTEC-mediated expression of antigens associated with peripheral tissues plays a pivotal role in the enforcement of central tolerance through the deletion of autoreactive T-cell clones. However, while Aire would appear to control the expression of a large array of peripheral tissue antigens, it is important to note that not all ectopic peripheral tissue antigen expression within mTEC is Aire-dependent (Anderson et al. 2002; Derbinski et al. 2005). Initial experiments have indicated that at a least a portion of Aire-independent antigens are influenced by signaling through the lymphotoxin pathway (Seach et al. 2008), it still remains to be determined precisely how the complete array of intrathymic TRA are regulated within the mTEC compartment.

A key question in the efficiency of thymic negative selection is posed by how large cohorts of developing T-cells are successfully screened by a relatively minor fraction of mTEC. Compounding this high ratio of thymocyte to mTEC distribution is the high selective distribution of any single given peripheral tissue antigen. In this regard, it has previously been estimated that any individual TRA is expressed by less than 5 % of total Aire-positive mTEC, which themselves comprise a minor fraction of total mTEC (Derbinski et al. 2001, 2008). In order to effectively delete autoreactive T-cell clones, several coordinated mechanisms appear to operate in order to ensure an imposition of central tolerance upon thymocytes. While mTEC are essential cellular production units for TRA a sharing of labor exists between mTEC and tDC in the presentation of self-antigen. Transfer of mTEC-derived antigen would appear to occur in a directional manner from mTEC to tDC interestingly including both intracellular and cell surface expressed antigen (Koble and Kyewski 2009). However, as yet the precise mechanism of how mTEC-derived self-antigen is transferred to DC for presentation to T-cells remains currently unknown. The absolute necessity of tDC in the contribution to negative selection is suggested both by conditional deletion of CD11c-positive cells via targeted diphtheria toxin susceptibility leading to fatal autoimmunity and absence of DC-MHC expression leading to inefficient transgenic T-cell deletion against an mTEC-associated neo-antigen (Gallegos and Bevan 2004; Ohnmacht et al. 2009). The contribution of DC to negative selection likely facilitates the spreading of particular self-antigens within thymic microenvironments, such antigen spread may particularly be of note in regard to recent studies indicating a rather anatomically restricted range of intramedullary T-cell migration (Le Borgne et al. 2009).

While transfer of antigen from mTEC to DC likely plays an important role in the spreading of peripheral tissue antigen and presentation to developing T-cells, direct presentation of antigen by mTEC additionally seems to shape the TCR repertoire. Recent experiments have demonstrated that negative selection of transgenic CD4 T-cells was impaired when MHC class II expression was selectively reduced on mTEC (Hinterberger et al. 2010). Such findings clearly imply a direct role of mTEC in autonomous presentation of antigen to T-cells, in addition to acting as a peripheral tissue antigen reservoir for co-operative transfer to tDC. Outstanding questions in relation to medullary enforcement of negative selection remain, particularly regarding the specific routes of peptide generation and loading of endogenous self-antigens into MHC II pathways in medullary thymic epithelial cells. In addition, it is interesting to note that the major fraction of thymic DC are comprised of the CD205-expressing subset, being peripherally-associated with a capacity to present exogenous antigen into both MHC class-II, and -I via cross-presentation pathways and being capable of tolerance induction (Bonifaz et al. 2002, 2004). Whether such a proportional makeup of tDC reflects a functional association with transfer of antigen from mTEC and presentation to both CD4 and CD8 T-cells remains unclear. Further, as compared to peripherally equivalent DC, thymic resident DC demonstrate an enhanced capacity for antigen cross-presentation and T-cell cross-priming in the absence of DC activating factors (Dresch et al. 2011), suggesting that thymic microenvironments may uniquely influence the efficiency of antigen presentation via as yet undefined cellular and molecular mechanisms.

4.2 Foxp3 Regulatory T-Cell Development

While it would appear clear that deletional tolerance mediated by thymic medullary cellular microenvironments is essential for the removal of newly generated autoreactive T-cell clones as described above, the effectiveness of such tolerizing mechanisms does not appear to be one hundred percent efficient. The leakiness in the efficiency of T-cell negative selection can be clearly revealed by murine models lacking T-regulatory cell (T-Reg) populations. In the absence of Foxp3-dependent T-Reg, autoreactive T-cell clones normally present in the peripheral T-cell pool become apparent, with their unopposed activation rapidly leading to the generation of catastrophic and fatal systemic autoimmunity (Kim et al. 2007). Early experimentation demonstrating fatal autoimmunity in neonatal mice, having undergone early stage thymectomy, among other data, presented initial evidence that thymic microenvironments were essential for the generation of suppressive CD4 T-cells (Josefowicz et al. 2012). Subsequently, the requirement for thymic microenvironments in development of T-Reg was found to strictly depend upon the selection of T-Reg by intrathymic self-antigen expression (Itoh et al. 1999; Jordan et al. 2001). The notion that TCR specificity may influence $Foxp3^+$ T-Reg generation combines several pieces of evidence, including the finding that TCR usage of conventional versus regulatory T-cells indicated partially differential specificity with a low degree of overlay, at least in the context of an experimentally limited TCR repertoire (Hsieh et al. 2004; Pacholczyk et al. 2006). The question subsequently arising from the proposition that T-Reg are developmentally selected by TCR specificity for cognate self-antigen is how potential self-reactivity leads to a T-Reg fate versus the induction of apoptosis via negative selection. The primary notion in this regard involves a role for the strength of TCR self-reactivity, such that the selection of T-Reg occurs at an intermediate level between the low degrees of self-peptide:self-MHC required for positive selection and the high level of self-reactivity driving negative selection (Liston and Rudensky 2007). Directly in support of this theory, experiments utilizing microRNA-mediated MHC class II suppression in mTEC, resulting in a quantitative reduction in antigen presentation, resulted in the enhanced induction of T-Reg and a corresponding decline in negative selection (Hinterberger et al. 2010), implying that avidity plays a determining role in thymic T-Reg development. Interestingly, evidence from Nurr77-GFP mice, where levels of GFP expression correlate with intensity of TCR signal strength, indicate that thymic T-Reg would appear to experience a higher level of stimulation through their TCR than Foxp3-negative conventional T-cells (Moran et al. 2011).

The precise developmental timing of T-Reg generation has led to multiple lines of experimentation. While the induction of T-Reg was proposed to occur within the CD4$^+$8$^+$ fraction, being associated with cortical thymic localization and cortical cellular interactions including cTEC (Bensinger et al. 2001; Liston et al. 2008), subsequent studies have disputed the developmental significance of Foxp3$^+$ T-Reg generated within cortically restricted CD4$^+$8$^+$ stages (Lee and Hsieh 2009), instead suggesting that T-Reg in the main are generated following the transition to a CD4 SP (SP) stage. The cellular interactions leading to Foxp3 Treg generation therefore likely follow CD4 SP transition into thymic medullary environments, as mediated by CCR7 guided migration (Ueno et al. 2004). While the defining hallmark of T-Reg can be considered to be Foxp3 expression, it has been previously discovered that Foxp3$^+$CD25$^+$ intrathymic T-Reg appear to be derived from a Foxp3$^-$CD25$^+$ sub-population encompassing T-Reg progenitors (pT-Reg), as demonstrated by precursor-product experiments analyzing development of pT-Reg in vivo (Lio and Hsieh 2008). Interestingly, Foxp3$^-$CD25$^+$ pT-Reg selected by TCR-directed interaction with self-antigen were subsequently found to develop independently of TCR stimulation following their initial specification (Lio and Hsieh 2008). While this subsequent developmental step is proposed to be TCR-independent, evidence points to a cytokine-dependence of T-Reg maturation beyond the initial Foxp3$^+$CD25$^+$ stage, including signaling through Il-2 and IL-15 (Burchill et al. 2007), however the precise involvement of TCR-independent signaling in directing Foxp3$^+$ T-Reg maturation versus maintenance and survival remains unclear.

The differential ability of thymic medullary resident APC to dictate T-Reg induction has formed the basis for several experimental studies attempting to identify the key players in this important process. Primarily, both mTEC and thymic DC would appear to be able to efficiently induce T-Reg generation, as indicated by experimental systems providing selective absence and restriction of antigen expression to either population of APC (Aschenbrenner et al. 2007; Proietto et al. 2008; Spence and Green 2008). Thus, the capacity to efficiently select T-Reg does not reside within any single thymic APC population. While the ability of medullary-resident APC would seem to be promiscuous in the ability to select Foxp3 T-Reg, the antigen array responsible for selecting such cells remains unclear. The relatively small zonal territory of T-cells in medullary epithelium and their propensity to demonstrate increased dwelling time with medullary APC following recognition of cognate antigen, in a transgenic TCR model (Le Borgne et al. 2009), may fit with the interesting finding that the efficiency of intrathymic T-Reg development is highly dependent upon competition of specific T-cell clones

Mechanisms of Thymus Medulla Development and Function

for selecting antigen (Bautista et al. 2009). Such a balance of competition for selecting ligand in relation to induction of a T-Reg fate may play a pivotal role in determining the frequency of developing T-cells undergoing either negative selection, T-Reg fate induction, or progression as a conventional naïve T-cell. Further, the finding that the frequency of given TCR clones within the T-cell compartment is required to be below 1 % for efficient Treg generation, at least in the context of monoclonal transgenic T-cells, (Bautista et al. 2009), may suggest that a major fraction of thymic T-Reg are specifically selected by infrequently expressed antigen. Such scarce selecting antigen may potentially reflect peripheral tissue antigens expressed by mTEC, presenting the possibility that thymically derived T-Reg may display preferential specificity toward defined tissue-associated antigens rather than broadly expressed ubiquitous self-antigen. However, it should be noted that recent experiments studying the role of thymic niche availability in the regulation of T-Reg generation in a polyclonal T-cell compartment have come to the opposing conclusion that niche availability does not limit T-Reg generation (Romagnoli et al. 2012). It therefore remains unclear precisely how the proportion of T-cells entering the T-Reg pathway is intrathymically regulated.

In addition to antigen presentation, provision of co-stimulation via the CD28:CD80/86 axis plays a key role in T-Reg generation, with an absence of CD28-mediated co-stimulation leading to a highly depleted T-Reg population (Tai et al. 2005). As expected from promiscuous T-Reg induction influenced by self-antigen presentation, again expression of CD80/86 on either mTEC or hemopoietic cells, including tDC, is equally able to induce T-Reg development (Roman et al. 2010). The ability of medullary APC subsets to influence T-Reg induction may therefore depend upon their ability to present self-antigen in conjunction with defined co-stimulation, rather than perhaps being absolutely dependent upon the provision of unique cell-specific signals or self-antigen arrays limited for instance solely to mTEC. While both mTEC and tDC are able to induce T-Reg generation in a quantitative fashion, it remains unclear whether any qualitative differences exist between mTEC versus tDC specified T-Reg. In this regard, it could be speculated that the spectrum of T-Reg clones induced by mTEC interaction may differ from those generated via extrathymically derived $CD8^-Sirpa^+$ tDC which are able to transport peripheral antigen, including blood-borne antigens, into the thymus (Baba et al. 2009; Li et al. 2009). Finally, while thymic DC are globally capable of efficient T-Reg induction, in vitro studies have proposed that differences may exist in the efficiency of extrathymically $CD8^-Sirpa^+$ and intrathymically generated $CD8^+Sirpa^-$ DC to induce T-Reg (Proietto et al. 2008). Of particular note, in an in vitro model, $CD8^-Sirpa^+$ DC were proposed to demonstrate a superior capacity to instruct T-Reg induction potentially by virtue of increased maturity phenotype, including MHC class II and CD80/86 expression levels, again suggesting that the ability of APC to induce signals through the TCR with a particular strength may link their ability to efficiently induce T-Reg. In addition, $CD8^-Sirpa^+$ DC are proposed to selectively produce the chemokines CCL17 and CCL22 compared to $CD8^+Sirpa^-$ DC, potentially enhancing their ability to interact with newly selected CD4 SP thymocytes bearing the cognate chemokine receptor CCR4 (Proietto et al. 2008). The correct localization and ability of medullary resident APC to efficiently interact with thymocytes is further highlighted by findings in mice lacking expression of the Aire-dependent chemokine XCL1. XCL1-deficient mice demonstrate aberrant intrathymic DC positioning, albeit at normal total numbers, and display a corresponding reduction in T-Reg development with associated autoimmune disease (Lei et al. 2011). Together such findings highlight that the correct anatomical organization and positioning of medullary thymic cellular components likely plays a key role in influencing the efficiency of T-Reg development.

4.3 Post-Selection Thymocyte Differentiation

Upon entry of newly selected T-cells into medullary microenvironments, a period of medullary residency is essential to ensure sufficient screening of $CD4^+$ and $CD8^+$ thymocytes potentially preventing the escape of autoreactive clones into the periphery. It would therefore appear logical that specific mechanisms may operate in order to ensure that maturing SP thymocytes are retained within thymic medulla for a sustained period of time. Analysis of SP thymocyte populations, particularly CD4 thymocytes, has clearly demonstrated a distinct series of phenotypic subsets proposed to reflect differential maturational states. Following positive selection CD4 SP thymocytes were initially described to demonstrate a heterogeneous mix of phenotypes, being primarily split into an immature and mature subset based on CD24 (heat-stable antigen) and Qa-2 expression (Ramsdell et al. 1991; Vicari et al. 1994; Wilson et al. 1988). While such immature and mature SP thymocyte subsets appeared to display differential responsiveness to external stimuli, including susceptibility to negative selection being associated with immature-type SP stages (Kishimoto and Sprent 1997), evidence of progressive maturation has only recently been directly presented. Specifically, four clearly defined subsets of CD4 thymocytes, termed SP1-4, defined as $CD69^+Qa2^-6C10^-$ (SP1), $CD69^+Qa2^-6C10^+$ (SP2), $CD69^-Qa2^-6C10^-$ (SP3), and $CD69^-Qa2^+6C10^-$ (SP4) were phenotypically identified in murine thymus. Direct in vivo injection of traceable SP1 CD4 thymocytes into adult murine thymus provided strong evidence for SP CD4 thymocyte maturation occurring in a regulated sequential fashion (Li et al. 2007).

Initial estimates of SP thymocyte dwell time within medullary microenvironments proposed a timespan in the region of 14 days (Egerton et al. 1990), although contrasting studies suggested that newly generated naïve thymocytes were found to emigrate following just 2 days post intrathymic BrdU labeling (Tough and Sprent 1994). A key question that related to this potential discrepancy in proposed length of medullary residency was how exit from thymic microenvironments was regulated. Two potential mechanisms proposed opposing models of either a random exit of SP thymocytes at multiple stages of maturation (lucky-dip) versus a linear, hierarchical mode whereby thymic exit was restricted to the most mature cells (Scollay and Godfrey 1995). Subsequent experiments utilizing a novel RAG2-GFP reporter mice, whereby GFP expression levels correlated with thymocyte maturation, directly demonstrated that SP thymocytes spent a relatively brief period of 4 days within thymic microenvironments prior to their export (McCaughtry et al. 2007). In addition, it was further substantiated that thymic egress was found to be limited to the most mature SP thymocytes (Li et al. 2007; McCaughtry et al. 2007). Such a relatively short time of SP thymocyte medullary habitation presumably equates to a highly efficient process of autoreactive T-cell screening against correspondingly rare cognate self-antigen, including peripheral tissue antigens. In addition, the window for negative selection would appear to be even shorter than the 4-day intramedullary window, assuming that susceptibility to negative selection is enhanced within immature SP CD4 thymocytes defined by $CD24^{hi}$ (Kishimoto and Sprent 1997), again further narrowing the time frame in which negative selection is effective (Weinreich and Hogquist 2008). In direct relation to the efficiency of thymocyte negative selection, titration experiments using reaggregate thymic organ culture techniques demonstrated that thymic DC are still able to mediate efficient negative selection even at 1 % of total cell numbers, emphasizing the efficiency of DC as potent mediators of negative selection (Anderson et al. 1998), potentially demonstrating the highly efficient nature whereby autoreactive T-cell clones can be screened within thymic medullary microenvironments. However, it is also possible that the relatively tight temporal availability of negative selection susceptibility may correspond with the potential escape of autoreactive T-cell clones into the peripheral repertoire as may occur in Treg-deficient mice (Kim et al. 2007). The extent to which autoreactive T-cells are able to escape negative selection in the adult steady-state thymus warrants further investigation, further, whether extended medullary dwell time of developing thymocytes at a negative selection susceptible stage could influence the efficiency of negative selection poses an additional point of interest. Of note, a recent study has demonstrated an increased intrathymic dwell time for newly generated T-Reg compared to conventional T-cells (Romagnoli et al. 2012). The mechanisms responsible for this discrepancy between these two related T-cell sublineages remain unknown, as does the functional significance, if any, of this phenomenon.

If maturation and exit of SP thymocytes is dependent upon a linear, hierarchical model, it follows that specific mechanisms must tightly regulate the selective ability of the most mature thymocytes to be released into the periphery. Indeed, the ability of thymocytes to exit thymus into the periphery was clearly shown to be highly dependent upon the action of the transcription factors Foxo1 and KLF2 (Bai et al. 2007; Carlson et al. 2006; Kerdiles et al. 2009) at least in part, regulating the expression of the cell surface receptor S1PR1 (Allende et al. 2004; Matloubian et al. 2004). Notably, expression of S1PR1 directs chemoattraction toward a gradient of S1P predominantly present in blood but also potentially produced by vascular endothelium and modulated by perivascular cells in the thymus leading to the observed exit of mature thymocytes at blood vessels located at the cortico-medullary junction (Pham et al. 2010; Zachariah and Cyster 2010). In direct correlation with the progressive maturation of SP1 > SP4 thymocytes, gradual increased expression of the previously mentioned factors, including S1PR1, has been recently reported (Teng et al. 2011). Of interest, while it seems apparent that the most mature SP thymocytes selectively exit the thymus, recent thymus emigrants still appear to require further maturation events in the periphery, as demonstrated by progressive Qa-2 upregulation and a notable decrease in proliferative response as compared to more mature naïve T-cells (Boursalian et al. 2004). Whether the developmental prompts facilitating such final maturation post-thymic exits are unique to peripheral environments or are shared with those of the thymic medulla is yet to be fully determined.

Assuming that ordered SP thymocyte maturation correspondingly determines the regulated exit of mature cells as described previously, it is of particular interest to determine whether this linear development occurs via thymocyte-autonomous process or is influenced by external medullary microenvironmental factors. Intrathymic injection of SP1 thymocytes into adult thymus has been observed to follow a maturation time of 2–3 days (Li et al. 2007), however it was also reported that a subset of mature SP4 thymocytes were found to reside within the host thymus for a period up to 7 days (Li et al. 2007), suggesting that while SP maturation correlates with a functional capacity to emigrate, additional influences extrinsic to thymocytes may impose upon the timing of thymocyte exit.

In relation to the question surrounding control of SP maturation, interesting in vitro data provided evidence that isolated immature SP1 thymocytes were capable of progression through maturation stages SP1 > SP2 > SP3 in the absence of additional cellular support, including mTEC, however IL-7 was found to be critical for the survival of such isolated cells in vitro (Li et al. 2007). However, investigation of this potential regulatory axis revealed that absence of functional IL-7 signaling in vivo did not lead to an impairment in the maturation of post-selection SP thymocytes (Weinreich et al. 2011), suggesting that while IL-7 is sufficient to facilitate SP survival it would not appear to be essential in order to drive differentiation and maturation. Further analysis of the thymic microenvironment-dependent developmental requirements for SP3 > SP4 transition identified a stage-specific requirement for medullary microenvironments as RelB-deficient mice (Burkly et al. 1995), that display a drastic impairment in mTEC generation, were found to lack SP3 > SP4 development (Li et al. 2007). In addition, analysis of Aire-deficient mice again revealed defective generation of the final stage of SP4 maturation. However, it remains unclear precisely how Aire[+] mTEC influence SP4 thymocyte maturation.

Aire expression, in addition to regulating ectopic peripheral antigen expression, also appears to control additional aspects of mTEC biology, including chemokine expression such as CCR7 and CCR4 ligands (Laan et al. 2009) potentially influencing the capacity of developing SP thymocytes to localize and interact with mTEC. While Aire regulates chemokines associated with the attraction of mature thymocytes, it also controls the expression of DC attractants including XCL1, which play an essential role in regulating the efficiency of DC-mediated thymocyte selection and interaction (Lei et al. 2011). Further, while RelB-deficient mice display mTEC-intrinsic defects, DC also demonstrate altered representation within thymic microenvironments both due to mTEC-deficiencies and due to a DC-intrinsic influence of RelB (Burkly et al. 1995; Wu et al. 1998). Further experiments investigating the complex cellular interplay required for regulated SP thymocyte maturation leading to efficient deletional tolerance, Treg induction, and appropriate thymocyte maturation leading to tightly regulated thymic export will provide valuable insights into our current understanding of how medullary thymic microenvironments function.

5 Conclusions

The thymic medulla represents a key site in intrathymic $\alpha\beta$T-cell development, by controlling the fate of thymocytes that have undergone positive selection in the thymic cortex. The induction of T-cell tolerance in the medulla is controlled by multiple mechanisms: central tolerance results in the elimination of autoreactive $\alpha\beta$TCR specificities, while the development of Foxp3$^+$ Regulatory T-cells ensures that dominant tolerance can occur within peripheral body tissues. An increasing body of evidence supports the idea that these mechanisms of T-cell tolerance require both medullary epithelial cells, including the Aire$^+$ subset, and thymic dendritic cells that act in concert to shape the developing T-cell receptor repertoire. While some aspects of thymus medulla function are beginning to be defined, further studies are required to investigate several key aspects that remain poorly understood, including the identity and requirements of mTEC progenitors, the role of Aire in thymus medulla organization and mTEC development, and the importance of mTEC in post-selection maturation of conventional $\alpha\beta$T-cells. Perhaps most importantly, how the medulla controls the balance between negative selection and T-Reg production that ultimately results in a self-tolerant state is poorly understood. Gaining a better understanding of these features of thymic medulla function should help in identifying the cellular and molecular basis of T-cell mediated autoimmunity, and could inform future therapeutic strategies aimed at its treatment. A recent study from our laboratory has shown that medullary thymic epithelial cells are essential for the development of Foxp3+ T-Reg but are not required for continued development of conventional CD4+ thymocytes (Cowan et al 2013)

References

Adkins B, Gandour D, Strober S, Weissman I (1988) Total lymphoid irradiation leads to transient depletion of the mouse thymic medulla and persistent abnormalities among medullary stromal cells. J Immunol 140:3373–3379

Akiyama T, Maeda S, Yamane S, Ogino K, Kasai M, Kajiura F, Matsumoto M, Inoue J (2005) Dependence of self-tolerance on TRAF6-directed development of thymic stroma. Science 308:248–251

Akiyama T, Shimo Y, Yanai H, Qin J, Ohshima D, Maruyama Y, Asaumi Y, Kitazawa J, Takayanagi H, Penninger JM, Matsumoto M, Nitta T, Takahama Y, Inoue J (2008) The tumor necrosis factor family receptors RANK and CD40 cooperatively establish the thymic medullary microenvironment and self-tolerance. Immunity 29:423–437

Allende ML, Sasaki T, Kawai H, Olivera A, Mi Y, van Echten-Deckert G, Hajdu R, Rosenbach M, Keohane CA, Mandala S, Spiegel S, Proia RL (2004) Mice deficient in sphingosine kinase 1 are rendered lymphopenic by FTY720. J Biol Chem 279:52487–52492

Alves NL, Huntington ND, Rodewald HR, Di Santo JP (2009) Thymic epithelial cells: the multi-tasking framework of the T cell "cradle". Trends Immunol 30:468–474

Anderson G, Takahama Y (2012) Thymic epithelial cells: working class heroes for T cell development and repertoire selection. Trends Immunol 33:256–263

Anderson G, Partington KM, Jenkinson EJ (1998) Differential effects of peptide diversity and stromal cell type in positive and negative selection in the thymus. J Immunol 161:6599–6603

Anderson M, Anderson SK, Farr AG (2000) Thymic vasculature: organizer of the medullary epithelial compartment? Int Immunol 12:1105–1110

Anderson MS, Venanzi ES, Klein L, Chen Z, Berzins SP, Turley SJ, von Boehmer H, Bronson R, Dierich A, Benoist C, Mathis D (2002) Projection of an immunological self shadow within the thymus by the aire protein. Science 298:1395–1401

Anderson G, Lane PJ, Jenkinson EJ (2007) Generating intrathymic microenvironments to establish T-cell tolerance. Nat Rev Immunol 7:954–963

Aschenbrenner K, D'Cruz LM, Vollmann EH, Hinterberger M, Emmerich J, Swee LK, Rolink A, Klein L (2007) Selection of Foxp3+ regulatory T cells specific for self antigen expressed and presented by Aire+ medullary thymic epithelial cells. Nat Immunol 8:351–358

Baba T, Nakamoto Y, Mukaida N (2009) Crucial contribution of thymic Sirp alpha+ conventional dendritic cells to central tolerance against blood-borne antigens in a CCR2-dependent manner. J Immunol 183:3053–3063

Bai A, Hu H, Yeung M, Chen J (2007) Kruppel-like factor 2 controls T cell trafficking by activating L-selectin (CD62L) and sphingosine-1-phosphate receptor 1 transcription. J Immunol 178:7632–7639

Bautista JL, Lio CW, Lathrop SK, Forbush K, Liang Y, Luo J, Rudensky AY, Hsieh CS (2009) Intraclonal competition limits the fate determination of regulatory T cells in the thymus. Nat Immunol 10:610–617

Bensinger SJ, Bandeira A, Jordan MS, Caton AJ, Laufer TM (2001) Major histocompatibility complex class II-positive cortical epithelium mediates the selection of CD4(+)25(+) immunoregulatory T cells. J Exp Med 194:427–438

Blackburn CC, Augustine CL, Li R, Harvey RP, Malin MA, Boyd RL, Miller JF, Morahan G (1996) The nu gene acts cell-autonomously and is required for differentiation of thymic epithelial progenitors. Proc Natl Acad Sci U S A 93:5742–5746

Bleul CC, Corbeaux T, Reuter A, Fisch P, Monting JS, Boehm T (2006) Formation of a functional thymus initiated by a postnatal epithelial progenitor cell. Nature 441:992–996

Boehm T, Scheu S, Pfeffer K, Bleul CC (2003) Thymic medullary epithelial cell differentiation, thymocyte emigration, and the control of autoimmunity require lympho-epithelial cross talk via LTbetaR. J Exp Med 198:757–769

Bonifaz L, Bonnyay D, Mahnke K, Rivera M, Nussenzweig MC, Steinman RM (2002) Efficient targeting of protein antigen to the dendritic cell receptor DEC-205 in the steady state leads to antigen presentation on major histocompatibility complex class I products and peripheral CD8+ T cell tolerance. J Exp Med 196:1627–1638

Bonifaz LC, Bonnyay DP, Charalambous A, Darguste DI, Fujii S, Soares H, Brimnes MK, Moltedo B, Moran TM, Steinman RM (2004) In vivo targeting of antigens to maturing dendritic cells via the DEC-205 receptor improves T cell vaccination. J Exp Med 199:815–824

Boursalian TE, Golob J, Soper DM, Cooper CJ, Fink PJ (2004) Continued maturation of thymic emigrants in the periphery. Nat Immunol 5:418–425

Burchill MA, Yang J, Vogtenhuber C, Blazar BR, Farrar MA (2007) IL-2 receptor beta-dependent STAT5 activation is required for the development of Foxp3+ regulatory T cells. J Immunol 178:280–290

Burkly L, Hession C, Ogata L, Reilly C, Marconi LA, Olson D, Tizard R, Cate R, Lo D (1995) Expression of relB is required for the development of thymic medulla and dendritic cells. Nature 373:531–536

Carlson CM, Endrizzi BT, Wu J, Ding X, Weinreich MA, Walsh ER, Wani MA, Lingrel JB, Hogquist KA, Jameson SC (2006) Kruppel-like factor 2 regulates thymocyte and T-cell migration. Nature 442:299–302

Chin RK, Lo JC, Kim O, Blink SE, Christiansen PA, Peterson P, Wang Y, Ware C, Fu YX (2003) Lymphotoxin pathway directs thymic Aire expression. Nat Immunol 4:1121–1127

Chin RK, Zhu M, Christiansen PA, Liu W, Ware C, Peltonen L, Zhang X, Guo L, Han S, Zheng B, Fu YX (2006) Lymphotoxin pathway-directed, autoimmune regulator-independent central tolerance to arthritogenic collagen. J Immunol 177:290–297

Cowan JE, Parnell SM, Nakamura K, Caamano JH, Lane PJ, Jenkinson EJ, Jenkinson WE, Anderson G (2013) The thymic medulla is required for Foxp3+ regulatory but not conventional CD4+ thymocyte development. J Exp Med 210:675–681

Cupedo T, Kraal G, Mebius RE (2002) The role of CD45+CD4+CD3− cells in lymphoid organ development. Immunol Rev 189:41–50

Derbinski J, Kyewski B (2005) Linking signalling pathways, thymic stroma integrity and autoimmunity. Trends Immunol 26:503–506

Derbinski J, Schulte A, Kyewski B, Klein L (2001) Promiscuous gene expression in medullary thymic epithelial cells mirrors the peripheral self. Nat Immunol 2:1032–1039

Derbinski J, Gabler J, Brors B, Tierling S, Jonnakuty S, Hergenhahn M, Peltonen L, Walter J, Kyewski B (2005) Promiscuous gene expression in thymic epithelial cells is regulated at multiple levels. J Exp Med 202:33–45

Derbinski J, Pinto S, Rosch S, Hexel K, Kyewski B (2008) Promiscuous gene expression patterns in single medullary thymic epithelial cells argue for a stochastic mechanism. Proc Natl Acad Sci U S A 105:657–662

Desanti GE, Cowan JE, Baik S, Parnell SM, White AJ, Penninger JM, Lane PJ, Jenkinson EJ, Jenkinson WE, Anderson G (2012) Developmentally regulated availability of RANKL and CD40 ligand reveals distinct mechanisms of fetal and adult cross-talk in the thymus medulla. J Immunol 189:5519–5526

DeVoss J, Hou Y, Johannes K, Lu W, Liou GI, Rinn J, Chang H, Caspi RR, Fong L, Anderson MS (2006) Spontaneous autoimmunity prevented by thymic expression of a single self-antigen. J Exp Med 203:2727–2735

Dooley J, Erickson M, Farr AG (2008) Alterations of the medullary epithelial compartment in the Aire-deficient thymus: Implications for programs of thymic epithelial differentiation. J Immunol 181:5225–5232

Dresch C, Ackermann M, Vogt B, de Andrade Pereira B, Shortman K, Fraefel C (2011) Thymic but not splenic CD8(+) DCs can efficiently cross-prime T cells in the absence of licensing factors. Eur J Immunol 41:2544–2555

Drumea-Mirancea M, Wessels JT, Muller CA, Essl M, Eble JA, Tolosa E, Koch M, Reinhardt DP, Sixt M, Sorokin L, Stierhof YD, Schwarz H, Klein G (2006) Characterization of a conduit system containing laminin-5 in the human thymus: a potential transport system for small molecules. J Cell Sci 119:1396–1405

Dudakov JA, Hanash AM, Jenq RR, Young LF, Ghosh A, Singer NV, West ML, Smith OM, Holland AM, Tsai JJ, Boyd RL, van den Brink MR (2012) Interleukin-22 drives endogenous thymic regeneration in mice. Science 336:91–95

Egerton M, Scollay R, Shortman K (1990) Kinetics of mature T-cell development in the thymus. Proc Natl Acad Sci U S A 87:2579–2582

Farr AG, Anderson SK (1985) Epithelial heterogeneity in the murine thymus: fucose-specific lectins bind medullary epithelial cells. J Immunol 134:2971–2977

Farr AG, Braddy SC (1989) Patterns of keratin expression in the murine thymus. Anat Rec 224:374–378

Foster K, Sheridan J, Veiga-Fernandes H, Roderick K, Pachnis V, Adams R, Blackburn C, Kioussis D, Coles M (2008) Contribution of neural crest-derived cells in the embryonic and adult thymus. J Immunol 180:3183–3189

Gabler J, Arnold J, Kyewski B (2007) Promiscuous gene expression and the developmental dynamics of medullary thymic epithelial cells. Eur J Immunol 37:3363–3372

Gallegos AM, Bevan MJ (2004) Central tolerance to tissue-specific antigens mediated by direct and indirect antigen presentation. J Exp Med 200:1039–1049

Gillard GO, Farr AG (2005) Contrasting models of promiscuous gene expression by thymic epithelium. J Exp Med 202:15–19

Gillard GO, Dooley J, Erickson M, Peltonen L, Farr AG (2007) Aire-dependent alterations in medullary thymic epithelium indicate a role for Aire in thymic epithelial differentiation. J Immunol 178:3007–3015

Godfrey DI, Izon DJ, Tucek CL, Wilson TJ, Boyd RL (1990) The phenotypic heterogeneity of mouse thymic stromal cells. Immunology 70:66–74

Gray DH, Abramson J, Benoist C, Mathis D (2007) Proliferative arrest and rapid turnover of thymic epithelial cells expressing Aire. J Exp Med 204:2521–2528

Gray DH, Seach N, Ueno T, Milton MK, Liston A, Lew AM, Goodnow CC, Boyd RL (2006) Developmental kinetics, turnover, and stimulatory capacity of thymic epithelial cells. Blood 108:3777–3785

Guerau-de-Arellano M, Martinic M, Benoist C, Mathis D (2009) Neonatal tolerance revisited: a perinatal window for Aire control of autoimmunity. J Exp Med 206:1245–1252

Hadeiba H, Lahl K, Edalati A, Oderup C, Habtezion A, Pachynski R, Nguyen L, Ghodsi A, Adler S, Butcher EC (2012) Plasmacytoid dendritic cells transport peripheral antigens to the thymus to promote central tolerance. Immunity 36:438–450

Hamazaki Y, Fujita H, Kobayashi T, Choi Y, Scott HS, Matsumoto M, Minato N (2007) Medullary thymic epithelial cells expressing Aire represent a unique lineage derived from cells expressing claudin. Nat Immunol 8:304–311

Heino M, Peterson P, Sillanpaa N, Guerin S, Wu L, Anderson G, Scott HS, Antonarakis SE, Kudoh J, Shimizu N, Jenkinson EJ, Naquet P, Krohn KJ (2000) RNA and protein expression of the murine autoimmune regulator gene (Aire) in normal, RelB-deficient and in NOD mouse. Eur J Immunol 30:1884–1893

Hikosaka Y, Nitta T, Ohigashi I, Yano K, Ishimaru N, Hayashi Y, Matsumoto M, Matsuo K, Penninger JM, Takayanagi H, Yokota Y, Yamada H, Yoshikai Y, Inoue J, Akiyama T, Takahama Y (2008) The cytokine RANKL produced by positively selected thymocytes fosters medullary thymic epithelial cells that express autoimmune regulator. Immunity 29:438–450

Hinterberger M, Aichinger M, Prazeres da Costa O, Voehringer D, Hoffmann R, Klein L (2010) Autonomous role of medullary thymic epithelial cells in central CD4(+) T cell tolerance. Nat Immunol 11:512–519

Hsieh CS, Liang Y, Tyznik AJ, Self SG, Liggitt D, Rudensky AY (2004) Recognition of the peripheral self by naturally arising CD25+ CD4+ T cell receptors. Immunity 21:267–277

Hubert FX, Kinkel SA, Webster KE, Cannon P, Crewther PE, Proeitto AI, Wu L, Heath WR, Scott HS (2008) A specific anti-Aire antibody reveals aire expression is restricted to medullary thymic epithelial cells and not expressed in periphery. J Immunol 180:3824–3832

Irla M, Hugues S, Gill J, Nitta T, Hikosaka Y, Williams IR, Hubert FX, Scott HS, Takahama Y, Hollander GA, Reith W (2008) Autoantigen-specific interactions with CD4+ thymocytes control mature medullary thymic epithelial cell cellularity. Immunity 29:451–463

Itoh M, Takahashi T, Sakaguchi N, Kuniyasu Y, Shimizu J, Otsuka F, Sakaguchi S (1999) Thymus and autoimmunity: production of CD25+CD4+ naturally anergic and suppressive T cells as a key function of the thymus in maintaining immunologic self-tolerance. J Immunol 162:5317–5326

Izon DJ, Nieland JD, Godfrey DI, Boyd RL, Kruisbeek AM (1994) Flow cytometric analysis reveals unexpected shared antigens between histologically defined populations of thymic stromal cells. Int Immunol 6:31–39

Jenkinson WE, Jenkinson EJ, Anderson G (2003) Differential requirement for mesenchyme in the proliferation and maturation of thymic epithelial progenitors. J Exp Med 198:325–332

Jenkinson WE, Rossi SW, Parnell SM, Jenkinson EJ, Anderson G (2007) PDGFRalpha-expressing mesenchyme regulates thymus growth and the availability of intrathymic niches. Blood 109:954–960

Jordan MS, Boesteanu A, Reed AJ, Petrone AL, Holenbeck AE, Lerman MA, Naji A, Caton AJ (2001) Thymic selection of CD4+CD25+ regulatory T cells induced by an agonist self-peptide. Nat Immunol 2:301–306

Josefowicz SZ, Lu LF, Rudensky AY (2012) Regulatory T cells: mechanisms of differentiation and function. Annu Rev Immunol 30:531–564

Kajiura F, Sun S, Nomura T, Izumi K, Ueno T, Bando Y, Kuroda N, Han H, Li Y, Matsushima A, Takahama Y, Sakaguchi S, Mitani T, Matsumoto M (2004) NF-kappa B-inducing kinase establishes self-tolerance in a thymic stroma-dependent manner. J Immunol 172:2067–2075

Kanariou M, Huby R, Ladyman H, Colic M, Sivolapenko G, Lampert I, Ritter M (1989) Immunosuppression with cyclosporin A alters the thymic microenvironment. Clin Exp Immunol 78:263–270

Kerdiles YM, Beisner DR, Tinoco R, Dejean AS, Castrillon DH, DePinho RA, Hedrick SM (2009) Foxo1 links homing and survival of naive T cells by regulating L-selectin, CCR7 and interleukin 7 receptor. Nat Immunol 10:176–184

Kim JM, Rasmussen JP, Rudensky AY (2007) Regulatory T cells prevent catastrophic autoimmunity throughout the lifespan of mice. Nat Immunol 8:191–197

Kishimoto H, Sprent J (1997) Negative selection in the thymus includes semimature T cells. J Exp Med 185:263–271

Klein L, Hinterberger M, von Rohrscheidt J, Aichinger M (2011) Autonomous versus dendritic cell-dependent contributions of medullary thymic epithelial cells to central tolerance. Trends Immunol 32:188–193

Klug DB, Carter C, Crouch E, Roop D, Conti CJ, Richie ER (1998) Interdependence of cortical thymic epithelial cell differentiation and T-lineage commitment. Proc Natl Acad Sci U S A 95:11822–11827

Klug DB, Carter C, Gimenez-Conti IB, Richie ER (2002) Thymocyte-independent and thymocyte-dependent phases of epithelial patterning in the fetal thymus. J Immunol 169:2842–2845

Koble C, Kyewski B (2009) The thymic medulla: a unique microenvironment for intercellular self-antigen transfer. J Exp Med 206:1505–1513

Laan M, Kisand K, Kont V, Moll K, Tserel L, Scott HS, Peterson P (2009) Autoimmune regulator deficiency results in decreased expression of CCR4 and CCR7 ligands and in delayed migration of CD4+ thymocytes. J Immunol 183:7682–7691

Ladi E, Schwickert TA, Chtanova T, Chen Y, Herzmark P, Yin X, Aaron H, Chan SW, Lipp M, Roysam B, Robey EA (2008) Thymocyte-dendritic cell interactions near sources of CCR7 ligands in the thymic cortex. J Immunol 181:7014–7023

Le Borgne M, Ladi E, Dzhagalov I, Herzmark P, Liao YF, Chakraborty AK, Robey EA (2009) The impact of negative selection on thymocyte migration in the medulla. Nat Immunol 10:823–830

Lee HM, Hsieh CS (2009) Rare development of Foxp3+ thymocytes in the CD4+CD8+ subset. J Immunol 183:2261–2266

Lei Y, Ripen AM, Ishimaru N, Ohigashi I, Nagasawa T, Jeker LT, Bosl MR, Hollander GA, Hayashi Y, Malefyt Rde W, Nitta T, Takahama Y (2011) Aire-dependent production of XCL1 mediates medullary accumulation of thymic dendritic cells and contributes to regulatory T cell development. J Exp Med 208:383–394

Li J, Li Y, Yao JY, Jin R, Zhu MZ, Qian XP, Zhang J, Fu YX, Wu L, Zhang Y, Chen WF (2007) Developmental pathway of CD4+CD8− medullary thymocytes during mouse ontogeny and its defect in Aire−/− mice. Proc Natl Acad Sci U S A 104:18175–18180

Li J, Park J, Foss D, Goldschneider I (2009) Thymus-homing peripheral dendritic cells constitute two of the three major subsets of dendritic cells in the steady-state thymus. J Exp Med 206:607–622

Lio CW, Hsieh CS (2008) A two-step process for thymic regulatory T cell development. Immunity 28:100–111

Liston A, Rudensky AY (2007) Thymic development and peripheral homeostasis of regulatory T cells. Curr Opin Immunol 19:176–185

Liston A, Lesage S, Wilson J, Peltonen L, Goodnow CC (2003) Aire regulates negative selection of organ-specific T cells. Nat Immunol 4:350–354

Liston A, Nutsch KM, Farr AG, Lund JM, Rasmussen JP, Koni PA, Rudensky AY (2008) Differentiation of regulatory Foxp3+ T cells in the thymic cortex. Proc Natl Acad Sci U S A 105:11903–11908

Manley NR, Blackburn CC (2003) A developmental look at thymus organogenesis: where do the non-hematopoietic cells in the thymus come from? Curr Opin Immunol 15:225–232

Martins VC, Boehm T, Bleul CC (2008) Ltbetar signaling does not regulate Aire-dependent transcripts in medullary thymic epithelial cells. J Immunol 181:400–407

Mathis D, Benoist C (2009) Aire. Annu Rev Immunol 27:287–312

Matloubian M, Lo CG, Cinamon G, Lesneski MJ, Xu Y, Brinkmann V, Allende ML, Proia RL, Cyster JG (2004) Lymphocyte egress from thymus and peripheral lymphoid organs is dependent on S1P receptor 1. Nature 427:355–360

Matsumoto M (2011) Contrasting models for the roles of Aire in the differentiation program of epithelial cells in the thymic medulla. Eur J Immunol 41:12–17

McCaughtry TM, Wilken MS, Hogquist KA (2007) Thymic emigration revisited. J Exp Med 204:2513–2520

Moran AE, Holzapfel KL, Xing Y, Cunningham NR, Maltzman JS, Punt J, Hogquist KA (2011) T cell receptor signal strength in Treg and iNKT cell development demonstrated by a novel fluorescent reporter mouse. J Exp Med 208:1279–1289

Mori K, Itoi M, Tsukamoto N, Kubo H, Amagai T (2007) The perivascular space as a path of hematopoietic progenitor cells and mature T cells between the blood circulation and the thymic parenchyma. Int Immunol 19:745–753

Muller SM, Terszowski G, Blum C, Haller C, Anquez V, Kuschert S, Carmeliet P, Augustin HG, Rodewald HR (2005) Gene targeting of VEGF-A in thymus epithelium disrupts thymus blood vessel architecture. Proc Natl Acad Sci U S A 102:10587–10592

Muller SM, Stolt CC, Terszowski G, Blum C, Amagai T, Kessaris N, Iannarelli P, Richardson WD, Wegner M, Rodewald HR (2008) Neural crest origin of perivascular mesenchyme in the adult thymus. J Immunol 180:5344–5351

Nagamine K, Peterson P, Scott HS, Kudoh J, Minoshima S, Heino M, Krohn KJ, Lalioti MD, Mullis PE, Antonarakis SE, Kawasaki K, Asakawa S, Ito F, Shimizu N (1997) Positional cloning of the APECED gene. Nat Genet 17:393–398

Nehls M, Pfeifer D, Schorpp M, Hedrich H, Boehm T (1994) New member of the winged-helix protein family disrupted in mouse and rat nude mutations. Nature 372:103–107

Nehls M, Kyewski B, Messerle M, Waldschutz R, Schuddekopf K, Smith AJ, Boehm T (1996) Two genetically separable steps in the differentiation of thymic epithelium. Science 272:886–889

Nelson AJ, Dunn RJ, Peach R, Aruffo A, Farr AG (1996) The murine homolog of human Ep-CAM, a homotypic adhesion molecule, is expressed by thymocytes and thymic epithelial cells. Eur J Immunol 26:401–408

Nishikawa Y, Hirota F, Yano M, Kitajima H, Miyazaki J, Kawamoto H, Mouri Y, Matsumoto M (2010) Biphasic Aire expression in early embryos and in medullary thymic epithelial cells before end-stage terminal differentiation. J Exp Med 207:963–971

Nitta T, Ohigashi I, Nakagawa Y, Takahama Y (2011) Cytokine crosstalk for thymic medulla formation. Curr Opin Immunol 23:190–197

Nowell CS, Bredenkamp N, Tetelin S, Jin X, Tischner C, Vaidya H, Sheridan JM, Stenhouse FH, Heussen R, Smith AJ, Blackburn CC (2011) Foxn1 regulates lineage progression in cortical and medullary thymic epithelial cells but is dispensable for medullary sublineage divergence. PLoS Genet 7:e1002348

Odaka C (2009) Localization of mesenchymal cells in adult mouse thymus: their abnormal distribution in mice with disorganization of thymic medullary epithelium. J Histochem Cytochem 57:373–382

Odaka C, Morisada T, Oike Y, Suda T (2006) Distribution of lymphatic vessels in mouse thymus: immunofluorescence analysis. Cell Tissue Res 325:13–22

Ohnmacht C, Pullner A, King SB, Drexler I, Meier S, Brocker T, Voehringer D (2009) Constitutive ablation of dendritic cells breaks self-tolerance of CD4 T cells and results in spontaneous fatal autoimmunity. J Exp Med 206:549–559

Pacholczyk R, Ignatowicz H, Kraj P, Ignatowicz L (2006) Origin and T cell receptor diversity of Foxp3+CD4+CD25+ T cells. Immunity 25:249–259

Palmer DB, Viney JL, Ritter MA, Hayday AC, Owen MJ (1993) Expression of the alpha beta T-cell receptor is necessary for the generation of the thymic medulla. Dev Immunol 3:175–179

Petrie HT, Zuniga-Pflucker JC (2007) Zoned out: functional mapping of stromal signaling microenvironments in the thymus. Annu Rev Immunol 25:649–679

Pham TH, Baluk P, Xu Y, Grigorova I, Bankovich AJ, Pappu R, Coughlin SR, McDonald DM, Schwab SR, Cyster JG (2010) Lymphatic endothelial cell sphingosine kinase activity is required for lymphocyte egress and lymphatic patterning. J Exp Med 207:17–27

Porritt HE, Gordon K, Petrie HT (2003) Kinetics of steady-state differentiation and mapping of intrathymic-signaling environments by stem cell transplantation in nonirradiated mice. J Exp Med 198:957–962

Proietto AI, van Dommelen S, Zhou P, Rizzitelli A, D'Amico A, Steptoe RJ, Naik SH, Lahoud MH, Liu Y, Zheng P, Shortman K, Wu L (2008) Dendritic cells in the thymus contribute to T-regulatory cell induction. Proc Natl Acad Sci U S A 105:19869–19874

Ramsdell F, Jenkins M, Dinh Q, Fowlkes BJ (1991) The majority of CD4+8− thymocytes are functionally immature. J Immunol 147:1779–1785

Ramsey C, Winqvist O, Puhakka L, Halonen M, Moro A, Kampe O, Eskelin P, Pelto-Huikko M, Peltonen L (2002) Aire deficient mice develop multiple features of APECED phenotype and show altered immune response. Hum Mol Genet 11:397–409

Ripen AM, Nitta T, Murata S, Tanaka K, Takahama Y (2011) Ontogeny of thymic cortical epithelial cells expressing the thymoproteasome subunit beta5t. Eur J Immunol 41:1278–1287

Roberts NA, White AJ, Jenkinson WE, Turchinovich G, Nakamura K, Withers DR, McConnell FM, Desanti GE, Benezech C, Parnell SM, Cunningham AF, Paolino M, Penninger JM, Simon AK, Nitta T, Ohigashi I, Takahama Y, Caamano JH, Hayday AC, Lane PJ, Jenkinson EJ, Anderson G (2012) Rank signaling links the development of invariant gammadelta T cell progenitors and Aire(+) medullary epithelium. Immunity 36:427–437

Rodewald HR (2008) Thymus organogenesis. Annu Rev Immunol 26:355–388

Rodewald HR, Paul S, Haller C, Bluethmann H, Blum C (2001) Thymus medulla consisting of epithelial islets each derived from a single progenitor. Nature 414:763–768

Romagnoli P, Dooley J, Enault G, Vicente R, Malissen B, Liston A, van Meerwijk JP (2012) The thymic niche does not limit development of the naturally diverse population of mouse regulatory T lymphocytes. J Immunol 189:3831–3837

Roman E, Shino H, Qin FX, Liu YJ (2010) Cutting edge: Hematopoietic-derived APCs select regulatory T cells in thymus. J Immunol 185:3819–3823

Rossi SW, Jenkinson WE, Anderson G, Jenkinson EJ (2006) Clonal analysis reveals a common progenitor for thymic cortical and medullary epithelium. Nature 441:988–991

Rossi SW, Kim MY, Leibbrandt A, Parnell SM, Jenkinson WE, Glanville SH, McConnell FM, Scott HS, Penninger JM, Jenkinson EJ, Lane PJ, Anderson G (2007) RANK signals from

CD4(+)3(−) inducer cells regulate development of Aire-expressing epithelial cells in the thymic medulla. J Exp Med 204:1267–1272

Scollay R, Godfrey DI (1995) Thymic emigration: conveyor belts or lucky dips? Immunol Today 16:268–273, discussion 273–274

Seach N, Ueno T, Fletcher AL, Lowen T, Mattesich M, Engwerda CR, Scott HS, Ware CF, Chidgey AP, Gray DH, Boyd RL (2008) The lymphotoxin pathway regulates Aire-independent expression of ectopic genes and chemokines in thymic stromal cells. J Immunol 180:5384–5392

Seach N, Wong K, Hammett M, Boyd RL, Chidgey AP (2012) Purified enzymes improve isolation and characterization of the adult thymic epithelium. J Immunol Methods 385:23–34

Shakib S, Desanti GE, Jenkinson WE, Parnell SM, Jenkinson EJ, Anderson G (2009) Checkpoints in the development of thymic cortical epithelial cells. J Immunol 182:130–137

Shores EW, Van Ewijk W, Singer A (1991) Disorganization and restoration of thymic medullary epithelial cells in T cell receptor-negative scid mice: evidence that receptor-bearing lymphocytes influence maturation of the thymic microenvironment. Eur J Immunol 21:1657–1661

Shores EW, Van Ewijk W, Singer A (1994) Maturation of medullary thymic epithelium requires thymocytes expressing fully assembled CD3-TCR complexes. Int Immunol 6:1393–1402

Sitnik KM, Kotarsky K, White AJ, Jenkinson WE, Anderson G, Agace WW (2012) Mesenchymal cells regulate retinoic acid receptor-dependent cortical thymic epithelial cell homeostasis. J Immunol 188:4801–4809

Spence PJ, Green EA (2008) Foxp3+ regulatory T cells promiscuously accept thymic signals critical for their development. Proc Natl Acad Sci U S A 105:973–978

Surh CD, Ernst B, Sprent J (1992) Growth of epithelial cells in the thymic medulla is under the control of mature T cells. J Exp Med 176:611–616

Tai X, Cowan M, Feigenbaum L, Singer A (2005) CD28 costimulation of developing thymocytes induces Foxp3 expression and regulatory T cell differentiation independently of interleukin 2. Nat Immunol 6:152–162

Takahama Y (2006) Journey through the thymus: stromal guides for T-cell development and selection. Nat Rev Immunol 6:127–135

Teng F, Zhou Y, Jin R, Chen Y, Pei X, Liu Y, Dong J, Wang W, Pang X, Qian X, Chen WF, Zhang Y, Ge Q (2011) The molecular signature underlying the thymic migration and maturation of TCRalphabeta+ CD4+ CD8 thymocytes. PLoS One 6:e25567

Tough DF, Sprent J (1994) Turnover of naive- and memory-phenotype T cells. J Exp Med 179:1127–1135

Ueno T, Saito F, Gray DH, Kuse S, Hieshima K, Nakano H, Kakiuchi T, Lipp M, Boyd RL, Takahama Y (2004) CCR7 signals are essential for cortex-medulla migration of developing thymocytes. J Exp Med 200:493–505

van Ewijk W, Shores EW, Singer A (1994) Crosstalk in the mouse thymus. Immunol Today 15:214–217

Van Vliet E, Melis M, Van Ewijk W (1984) Monoclonal antibodies to stromal cell types of the mouse thymus. Eur J Immunol 14:524–529

Venanzi ES, Gray DH, Benoist C, Mathis D (2007) Lymphotoxin pathway and Aire influences on thymic medullary epithelial cells are unconnected. J Immunol 179:5693–5700

Vicari A, Abehsira-Amar O, Papiernik M, Boyd RL, Tucek CL (1994) MTS-32 monoclonal antibody defines CD4+8− thymocyte subsets that differ in their maturation level, lymphokine secretion, and selection patterns. J Immunol 152:2207–2213

Wang X, Laan M, Bichele R, Kisand K, Scott HS, Peterson P (2012) Post-Aire maturation of thymic medullary epithelial cells involves selective expression of keratinocyte-specific autoantigens. Front Immunol 3:19

Weih F, Caamano J (2003) Regulation of secondary lymphoid organ development by the nuclear factor-kappaB signal transduction pathway. Immunol Rev 195:91–105

Weih F, Carrasco D, Durham SK, Barton DS, Rizzo CA, Ryseck RP, Lira SA, Bravo R (1995) Multiorgan inflammation and hematopoietic abnormalities in mice with a targeted disruption of RelB, a member of the NF-kappa B/Rel family. Cell 80:331–340

Weinreich MA, Hogquist KA (2008) Thymic emigration: when and how T cells leave home. J Immunol 181:2265–2270

Weinreich MA, Jameson SC, Hogquist KA (2011) Postselection thymocyte maturation and emigration are independent of IL-7 and ERK5. J Immunol 186:1343–1347

White AJ, Withers DR, Parnell SM, Scott HS, Finke D, Lane PJ, Jenkinson EJ, Anderson G (2008) Sequential phases in the development of Aire-expressing medullary thymic epithelial cells involve distinct-cellular input. Eur J Immunol 38:942–947

White AJ, Nakamura K, Jenkinson WE, Saini M, Sinclair C, Seddon B, Narendran P, Pfeffer K, Nitta T, Takahama Y, Caamano JH, Lane PJ, Jenkinson EJ, Anderson G (2010) Lymphotoxin signals from positively selected thymocytes regulate the terminal differentiation of medullary thymic epithelial cells. J Immunol 185:4769–4776

Wilson A, Day LM, Scollay R, Shortman K (1988) Subpopulations of mature murine thymocytes: properties of CD4−CD8+ and CD4+CD8− thymocytes lacking the heat-stable antigen. Cell Immunol 117:312–326

Wu L, D'Amico A, Winkel KD, Suter M, Lo D, Shortman K (1998) RelB is essential for the development of myeloid-related CD8alpha− dendritic cells but not of lymphoid-related CD8alpha+ dendritic cells. Immunity 9:839–847

Yamazaki H, Sakata E, Yamane T, Yanagisawa A, Abe K, Yamamura KI, Hayashi SI, Kunisada T (2005) Presence and distribution of neural crest-derived cells in the murine developing thymus and their potential for differentiation. Int Immunol 17:549–558

Yano M, Kuroda N, Han H, Meguro-Horike M, Nishikawa Y, Kiyonari H, Maemura K, Yanagawa Y, Obata K, Takahashi S, Ikawa T, Satoh R, Kawamoto H, Mouri Y, Matsumoto M (2008) Aire controls the differentiation program of thymic epithelial cells in the medulla for the establishment of self-tolerance. J Exp Med 205:2827–2838

Zachariah MA, Cyster JG (2010) Neural crest-derived pericytes promote egress of mature thymocytes at the corticomedullary junction. Science 328:1129–1135

Zhu M, Chin RK, Tumanov AV, Liu X, Fu YX (2007) Lymphotoxin beta receptor is required for the migration and selection of autoreactive T cells in thymic medulla. J Immunol 179:8069–8075

Zuklys S, Balciunaite G, Agarwal A, Fasler-Kan E, Palmer E, Hollander GA (2000) Normal thymic architecture and negative selection are associated with Aire expression, the gene defective in the autoimmune-polyendocrinopathy-candidiasis-ectodermal dystrophy (APECED). J Immunol 165:1976–1983

Self-Peptides in TCR Repertoire Selection and Peripheral T Cell Function

Wan-Lin Lo and Paul M. Allen

Abstract The vertebrate antigen receptors are anticipatory in their antigen recognition and display a vast diversity. Antigen receptors are assembled through V(D)J recombination, in which one of each Variable, (Diverse), and Joining gene segment are randomly utilized and recombined. Both gene rearrangement and mutational insertion are generated through randomness; therefore, the process of antigen receptors generation requires a rigorous testing system to select every receptor which is useful to recognize foreign antigens, but which would cause no harm to self cells. In the case of T cell receptors (TCR), such a quality control responsibility rests in thymic positive and negative selection. In this review, we focus on the critical involvement of self-peptides in the generation of a T cell repertoire, discuss the role of T cell thymic development in shaping the specificity of TCR repertoire, and directing function fitness of mature T cells in periphery. Here, we consider thymic positive selection to be not merely a one-time maturing experience for an individual T cell, but a life-long imprinting which influences the function of each individual T cell in periphery.

Contents

1	The Role of Self-Peptides in Positive Selection	50
	1.1 Differential Strength of TCR Interaction with Positively Selecting Self-Peptides/MHC Instructs the Cell Fate of Double Positive Thymocytes	50
	1.2 Preselection DP Thymocytes are More Sensitive to Positively Selecting Self-Peptides/MHCthan Mature T Cells	51
	1.3 Self-Peptides Presented by Cortical Thymic Epithelial Cells are Essential for Positive Selection	53
2	The Specificity of Positively SelectingSelf-Peptide Recognition	54
	2.1 One Positively Selecting Self-peptide has to Select More than One TCR	54

W.-L. Lo · P. M. Allen (✉)
Department of Pathology and Immunology, Washington University School of Medicine, 660 S. Euclid, Box 8118, St. Louis, MO 63110, USA
e-mail: pallen@wustl.edu

Current Topics in Microbiology and Immunology (2014) 373: 49–67
DOI: 10.1007/82_2013_319
© Springer-Verlag Berlin Heidelberg 2013
Published Online: 24 April 2013

2.2	Recognition of Positively Selecting Ligand by a Given TCR Exhibits a High Degree of Specificity	54
2.3	Positively Selecting Ligand Shapes the Antigen Specificity of Post-Selection TCR Repertoire	56
3	The Role of Self-Peptides in Periphery	58
3.1	Positively Selecting Self-Peptides May Maintain Homeostatic Proliferation and Survivalof Peripheral T Cells	58
3.2	Positively Selecting Self-Peptides May Functionas Co-agonists to Augment FunctionalSensitivity of Peripheral T Cells	60
3.3	Positive Selection Signals May Program the Long-Term Survival and Function of Peripheral T Cells	61
4	Conclusion	63
References		63

1 The Role of Self-Peptides in Positive Selection

1.1 Differential Strength of TCR Interaction with Positively Selecting Self-Peptides/MHC Instructs the Cell Fate of Double Positive Thymocytes

Classical $\alpha\beta$ T cells differentiate from bone marrow-derived early thymic progenitors, through two developmental checkpoints: β selection at the double negative stage (DN), and positive and negative selections at double positive (DP) stage (Jameson et al. 1995; Moran and Hogquist 2012; Morris and Allen 2012). Thymocyte survival and lineage commitment require a TCR on a DP thymocyte to interact with peptide–MHC ligand on epithelial cells in the cortex. Only a weak interaction may support DP thymocytes to complete positive selection to become mature T cells, while the interaction either too strong or too weak would lead to negative selection or death by neglect, respectively. How can the same TCR initiate either a transcriptional program of survival and differentiation or that of death? The profiles of Ca^{2+} responses and ERK activation are critical for these outcomes. A transient, high-intensity burst of Ca^{2+} influx and ERK activation leads to negative selection, whereas positive selection requires sustained Ca^{2+} and ERK signaling (Mariathasan et al. 2001; McNeil et al. 2005; Werlen et al. 2000) (also see reviews in Moran and Hogquist 2012; Morris and Allen 2012).

The TCR and self-peptide-MHC interaction can occur over a wide range of affinities, raising the question as to how positive selection signals can trigger distinct genetic programs to initiate DP thymocytes to commit to one specific development pathway rather than another. Such a "decision" is taken according to the strength of positive selection signals, in conjunction with signals provided by the coreceptors. Among the affinity range of TCR:peptide MHC interaction, positive selection of $CD4^+$ T cells requires a stronger interaction than the selection of $CD8^+$ T cells. Strong and sustained TCR signals promote $CD4^+$ differentiation, whereas weaker and shorter signals generate $CD8^+$ T cells (Moran and Hogquist

2012). An even stronger signal than the one for CD4$^+$ selection may induce the development of regulatory T cells or innate-like T cells, such as natural killer T cells, and CD8$\alpha\alpha$ intestinal intraepithelial T cells (Stritesky et al. 2012).

After receiving positive selection signals, preselection DP cells downregulate CD8 coreceptors. If a DP cell recognizes MHC class II molecules, a stronger TCR signal promotes CD4$^+$ T cell development regardless of the downregulation of coreceptor CD8. However, if a preselection DP T cell recognizes MHC class I, the downregulation of coreceptor CD8 would cause TCR signaling to cease, so that CD8 coreceptor could not provide a signal strong enough to initiate a CD4$^+$ T cell differentiation, but instead advocate co-receptor reversal to support the development of CD8$^+$ T cells. Thus, the development of CD4$^+$ and CD8$^+$ T cells exhibits temporal difference that CD4$^+$ T cells can undergo rapid selection within 48 h, while CD8$^+$ T cells appear 4 days or more later (Saini et al. 2010). Several signaling molecules and transcription factors would lead to exclusive differentiation of either CD4$^+$ or CD8$^+$ T cell lineage (Rothenberg et al. 2008; Singer et al. 2008; Wang and Bosselut 2009). For example, ThPOK, or GATA-3, selectively promotes CD4$^+$ T cell differentiation, whereas Runx3 or Runx1 selectively induces CD8$^+$ T cells differentiation. Also, CD4 coreceptors have higher affinity for Lck than does CD8, endowing CD4 coreceptors to serve as better recruiters of Lck than CD8 (Alarcon and van Santen 2010; Hernandez-Hoyos et al. 2000; Legname et al. 2000; Schmedt et al. 1998; Wiest et al. 1993).

1.2 Preselection DP Thymocytes are More Sensitive to Positively Selecting Self-Peptides/MHC than Mature T Cells

Preselection DP thymocytes are more sensitive to activation than mature T cells (Davey et al. 1998; Eck et al. 2006; Sebzda et al. 1996; Stephen et al. 2009), despite the average number of TCRs being 10-fold lower than mature T cells (Bluestone et al. 1987; Havran et al. 1987). Positively selecting ligands can induce Ca^{2+} flux and upregulate CD69 expression on preselection DP thymocytes but not extrathymic mature T cells (Davey et al. 1998; Eck et al. 2006; Lo et al. 2009, 2012; Sebzda et al. 1996; Stephen et al. 2009). Preselection DP thymocytes are more responsive to low affinity positively selecting ligands, because of the lower expression of inhibitory coreceptors such as CD5 and CD45, lower expression of negative regulators in signaling pathways (such as tyrosine phosphatase SHP-1 and Cbl-b), and altered glycosylation of cell surface receptors (Azzam et al. 1998; McNeill et al. 2007; Plas et al. 1999; Chiang et al. 2000; Naramura et al. 1998; Davey et al. 1998). Moreover, preselection DP thymocytes also express stage-specific molecules that endow the sensitivity of DP thymocytes, including Tespa1 (which recruits PLC-γ1 and Grb2) (Wang et al. 2012), voltage-gated Na$^+$ channels (VGSC; composed of an a pore unit and a b regulatory subunit) (Lo et al. 2012), and miR181a (Ebert et al. 2009, 2010; Li et al. 2007).

The VGSC is composed of a pore-forming SCN5A subunit, and a regulatory SCN4B subunit. The expression of both subunits are tightly regulated that only DN3 (ready for β selection) and DP thymocytes (ready for positive selection) express *Scn5a* and *Scn4b*, but not mature T cells. Also, only positive selection signals can maintain the expression of *Scn5a* and *Scn4b* transcripts up to 7 h, in that negative selection signals downregulate both transcripts within 1 h. Blocking the SCN5A pore activity with a specific inhibitor tetrodotoxin diminished gp250-induced Ca^{2+} influx and has little effect on MCC-induced Ca^{2+} responses. In reaggregate culture, tetrodotoxin inhibited positive selection of AND $CD4^+$ T cells; in lentiviral bone marrow reconstitution experiments, shRNA knockdown of *Scn5a* inhibited thymic selection of $CD4^+$ T cells, but not $CD8^+$ T cells. The same $CD4^+$ selection specific defect was observed in in vitro reaggregate culture system. Peripheral AND $CD4^+$ T cells transfected with human SCN5A and SCN4B, which they normally do not express, gained the ability to upregulate CD69 expression in response to positively selecting ligands gp250, directly demonstrating that expression of VGSC contributes the increased sensitivity of DP thymocytes to weak positively selecting signals.

miR-181a is an microRNA that can enhance DP thymocyte sensitivity (Li et al. 2007) (Ebert et al. 2009, 2010). Like *Scn5a* and *Scn4b*, the expression of miR-181a precisely correlates with two selection events in thymocyte development: DN3 (β-selection) and DP (positive selection). Moreover, the expression of miR-181a is regulated by the strength of TCR signals, that the stronger the TCR signals (such as negative selection signals), the fewer miR-181a remains expressed after positive selection. miR-181a increases DP thymocyte sensitivity by inhibiting several phosphatases that negatively regulate the TCR signaling cascade, including Ptpn22 (which inhibits the phosphorylation of ZAP70 and Lck), Shp2 (a tyrosine kinase phosphatase), DUSP5, and DUSP6 (which inhibits ERK phosphorylation) (Ebert et al. 2009, 2010; Li et al. 2007). By negatively regulating the inhibitors in the TCR proximal signaling and ERK activation pathways, miR181-a therefore can enhance the sensitivity of DP thymocytes. Indeed, overexpression of miR-181a in mature T cells endows them to respond to ligands that were normally too weak to stimulate a response. Interestingly, similar to inhibition of VGSC pore function, inhibition of miR-181a in thymic cultures impaired Gag-Pol ability to positively select 5C.C7 $CD4^+$ T cells; however, miR-181a deficiency also disrupts the central tolerance that more self-reactive T cells were positively selected.

SCN5A-SCN4B composed VGSC and miR-181a share many similar features: the expression of both are very developmental stage-specific (only at DN3 and DP stages) and quickly regulated by the strength of TCR signaling; both enhance the sensitivity of preselection DP thymocytes to respond to weak ligands, and while ectopic expression in mature T cells, mature T cells acquire the ability to respond to the ligands that they normally do not respond; both play roles in the Ca^{2+} signaling pathways. Together both components endow preselection DP thymocytes to translate low affinity positively selecting ligand interactions to a series of strong, sustained Ca^{2+} flux and ERK activation to deliver a selection signal whose strength and duration is "just right" to initiate genetic programs required for T cell maturation but not to induce programmed cell death.

1.3 Self-Peptides Presented by Cortical Thymic Epithelial Cells are Essential for Positive Selection

Self-peptides presented by MHC molecules on cortical thymic epithelial cells (cTECs) in the thymus are a *sine qua non* for positive selection. The issue if thymic epithelial cells present a unique repertoire of self-peptides has been an active area of investigation for several decades (Lorenz and Allen 1988; Marrack et al. 1993), but little progress had been made in recent years until the unique protein degradation machinery was identified in cTEC. Importantly, cTECs uniquely express β5t-containing thymoproteasomes to process antigens for MHC class I-restricted presentation (see review in Takahama et al. 2010). The unique β5t subunit in the thymoproteasome favors the production of peptides that are less stably bound to MHC class I molecules, because β5t subunit does not efficiently cleave substrates at hydrophobic residues to generate optimal MHC class I binding peptides (Murata et al. 2007). This feature of β5t may be parallel to the observation that short-lived TCR engagement would promote the positive selection of CD8$^+$ T cell (Nitta et al. 2010). The deficiency of β5t subunit affected positive selection of CD8$^+$ T cells of MHC class I-restricted TCR transgenic mouse strains to different extents. The β5t deficiency decreased positive selection of HY TCR transgenic CD8$^+$ T cells strongly, but had no effect on OT-I TCR transgenic CD8$^+$ T cells (Nitta et al. 2010). These data suggested that some of the self-peptides displayed by cTEC are unique, while some of the self-peptides overlapped to those presented by peripheral antigen-presenting cells (APCs). The thymoproteasome-specific self-peptides are critical for the positive selection of most CD8$^+$ T cells (Nitta et al. 2010), and essential for the generation of an immunocompetent repertoire of CD8$^+$ T cells. However, thymoproteasome is not the sole component on cTEC to generate self-peptides for MHC class I-restricted positive selection.

As for peptide generation for MHC class II, cTECs express lysosomal cysteine proteases cathepsin L, but not cathepsin S as in B cells and dendritic cells (Honey and Rudensky 2003). Cathepsin L-deficient mice have impaired CD4$^+$ T cell selection, suggesting its critical role in generating self-peptides presented by MHC class II molecules (Honey et al. 2002; Nakagawa et al. 1998). However, cathepsin L expression is not exclusively restricted to cTECs, as macrophages also express cathepsin L (Hsieh et al. 2002; Nakagawa et al. 1998). While engineering a fibroblast cell line to express either cathepsin S or cathepsin L, and using mass spectrometry to analyze the study showed that self-peptides eluted from the MHC class II molecules of the fibroblasts, the overlap between the two self-peptide repertoires was substantial (Hsieh et al. 2002). Therefore, despite the expression of cathepsin L by cTECs and cathepsin S by B cells and dendritic cells, the universe of self-peptides presented by MHC class II on cTEC greatly overlaps with that presented by peripheral antigen-presenting cells. Thus, although cTECs appear to express some unique protein processing components, this has not been shown to result in the presentation of a truly unique repertoire of peptides.

2 The Specificity of Positively Selecting Self-Peptide Recognition

2.1 One Positively Selecting Self-peptide has to Select More than One TCR

Positively selecting self-peptides may directly influence the post-selection repertoire of mature T cells. For example, altered positively selecting self-peptides because of the β5t deficiency would affect the post-selection repertoire of mature T cells (Nitta et al. 2010). Such observations indicate DP thymocytes recognize positively selecting ligands with a certain degree of specificity, and raise two questions: first, to what degree of specificity do DP thymocytes recognize positively selecting ligands? Second, how does the specificity of positively selecting self-peptides affect the repertoire of mature T cells?

To examine the relationship between TCRs and a positively selecting self-peptide/MHC, a simple calculation reveals that it is not a monogamous relationship. A single positively selecting self-peptide must select multiple TCRs. For MHC class II-associated peptides, a B cell line expressing I-A^{g7} MHC class II molecules was used to isolate endogenously processed peptides. Mass spectrometry analysis estimated an individual APC could display roughly 2,000 different peptide families (a set of peptides with the same core P1–P9 residues but containing variable numbers of extensions at amino- and carboxy-terminus is defined as a peptide family) (Suri et al. 2002). Also, MHC I-associated peptides from EL4 thymoma cell lines were analyzed by mass spectrometry, compared with those from β2m-deficient EL4 mutant cells (Fortier et al. 2008). The data suggested thousands of peptides were present in low copy numbers per cell (Fortier et al. 2008). Given that a mouse is estimated to have a total of 30 million different T cells (Casrouge et al. 2000), one single peptide family must select several thousand T cells to generate a complete T cell repertoire (Fig. 1a). Even if we have underestimated the number of peptide families presented by cortical thymic epithelia cells by 10 or 100 fold due to the presence of many low-abundant self-peptides not detectable by current mass spectrometry techniques, one self-peptide-MHC still has to positively select more than one TCR.

2.2 Recognition of Positively Selecting Ligand by a Given TCR Exhibits a High Degree of Specificity

To directly examine the relationship between positively selecting self-peptides and post-selection TCR repertoire, we and others have identified naturally occurring positively selecting self-peptides for individual TCRs in the universe of self-peptides in vitro. The first naturally occurring positive selecting self-peptides

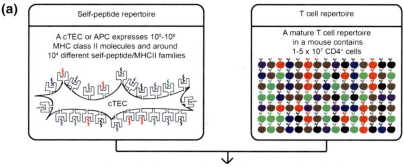

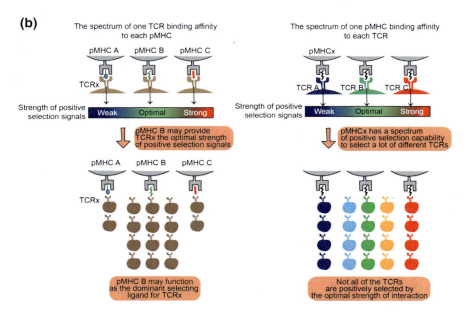

Fig. 1 a Illustration of the relation between self-peptide repertoire and T cell repertoire. **b** Illustration of the relation between positively selecting ligands and the TCRs. One individual TCR can be positively selected by more than one positively selecting ligand; however, for some TCRs, a dominant positively selecting ligand may exist, if there is a particular ligand that can provide the most optimal strength of pMHC/TCR interaction. Also, each selecting ligand has its own spectrum of positively selecting capability. One positively selecting ligand can select many different TCRs, but not all of the post-selection T cells are positively selected by the optimal strength of pMHC/TCR interaction

identified were the MHC class I-restricted Cappa1 peptide and β-catenin peptide for OT-I TCRs (Hogquist et al. 1997; Santori et al. 2002). The approach used was to screen self-peptides that were eluted from purified K^b molecules from EL4 T cell thymoma, LB27.4 B cell lymphoma cells, and thymi from C57BL/6 mice. These eluted self-peptides were fractionized by reverse phase HPLC and used in a coreceptor downregulation assay with TAP-I deficient OT-I TCR transgenic thymocytes. In 80 fractions, only two self-peptides, Cappa1 and β-catenin, were able to positively select OT-I TCR transgenic DP thymocytes, and none were able to positively select a closely related 2C TCR DP thymocytes (Hogquist et al. 1997; Santori et al. 2002).

For MHC class II, a panel of 95 I-E^k self-peptides eluted from CH27 mouse B cell lines has been screened for the ability of individual peptide to induce positive selection of five I-E^k-restricted TCR transgenic mouse lines: AND, 2.102, N3L2, A1, and 5C.C7 (Ebert et al. 2009; Lo et al. 2009). Among the 95 tested, only one peptide, gp250, was confirmed for its ability to positively select AND TCR (Lo et al. 2009), whereas for 5C.C7, six peptides were found in which Gag-Pol was the most potent one (Ebert et al. 2009). Among the same pool of 95 peptides, none were found to be able to positively select 2.102, N3L2, or A1 TCR (Lo et al. 2009). Such a frequency and low success rate in identifying a naturally occurring positively selecting ligand for MHCI- and MCHII-restricted TCRs imply that the recognition of positively selecting ligand has a certain degree of specificity. The recognition of positively selecting ligand with a high degree of specificity is further supported by experiments that a single amino acid substitution of gp250 disrupts its ability to positively select AND TCR. Moreover, the positive selection of the two highly similar TCRs, AND, and 5C.C7, are mediated by two mutually exclusive self-peptides. Both TCRs recognize agonist peptide MCC, uses the same TCR α and β segments (Vα11 and Vβ3), and only differ by four amino acids in CDR3α regions (Malherbe et al. 2004). But positively selecting peptide gp250 for AND TCR did not positively select 5C.C7 TCR, and selecting ligand Gag-Pol for 5C.C7 was unable to positively select AND TCR. Therefore, in vitro, preselection DP thymocytes recognize positively selecting ligands with a high degree of specificity. Whether recognition of positively selecting self-peptides in vivo requires a similarly high degree of specificity would require further studies.

2.3 Positively Selecting Ligand Shapes the Antigen Specificity of Post-Selection TCR Repertoire

One positively selecting self-peptide has to select at least several thousands of TCRs, and TCR recognition of positively selecting self-peptide/MHC complexes has a high degree of specificity in vitro. Does such a specific recognition during positive selection affect peptide specificity of the peripheral T cell repertoire? Several studies examined the question by starting with a defined positively selecting self-peptide, and examined the question that how many and what TCRs were positively selected by a single selecting ligand. Initial studies were performed on mouse strains engineered to express MHC class II complexes loaded with a single peptide, by either generating a H-2M deficient mouse line or introducing a transgene that covalently linked MHC to a specific peptide.

In H-2M deficient mice, the peptide exchange machinery is disrupted and MHC class II molecules are predominately bound by a single peptide species, CLIP. Approximately 30–50 % of normal numbers of CD4$^+$ T cells developed in H-2M deficient mice, and these post-selection CD4$^+$ T cells expressed full range of Vβ segments (Grubin et al. 1997; Surh et al. 1997). A similar phenotype was observed in Eα52-68/I-A^b single chain mice, in which a transgene encoding the Eα52-68 peptide covalently bound to the I-Aβ^b chain was introduced to invariant chain and endogenous I-Aβ^b deficient background (Ignatowicz et al. 1996, 1997). In Eα52-68/I-A^b single chain mice, the percentage of mature CD4$^+$ T cells decreased to around 20 % of wide type, and contained a full range of TCRβ usage (Ignatowicz et al. 1996, 1997). Both H-2M deficient and Eα52-68/I-A^b mice are capable of responding to immunization with multiple peptide antigens (Grubin et al. 1997; Ignatowicz et al. 1996, 1997; Surh et al. 1997). These studies demonstrated that a single predominant MHC bound peptide can positively select a very large and diverse repertoire of TCRs. However, these single peptide mice have skewed the amino acid frequency in the TCRα CDR3 loop and TCR Vα usage in positively selected T cells, and failed to positively select several defined clones of TCRs when introducing various transgenic TCRs onto these single peptide mice (Chmielowski et al. 2000; Fukui et al. 1997, 1998; Gapin et al. 1998; Grubin et al. 1997; Ignatowicz et al. 1996, 1997; Surh et al. 1997). Further studies showed that other low abundance self-peptides, not just the engineered dominant peptide, probably contributed to the generation of majority of post-selection TCR repertoire (Barton and Rudensky 1999). The study used mice expressing a human invariant chain (Ii) transgene in which CLIP region of human Ii was replaced with the Eα52-68 peptide (Ii-Eα mice). The Ii-Eα mice successfully restored the MHC class II expression to wild-type levels, but only 95 % of MHC class II molecules were bound with Eα52-68 peptide. These 5 % non-Eα52-68 peptides were dependent on H-2M molecules. When the H-2M deficiency was introduced to Ii-Eα mice, the number of CD4$^+$ T cells decreased to 30 % of that seen in H-2M sufficient Ii-Eα mice (Barton and Rudensky 1999). The data suggested both high and low-abundant self-peptides may contribute to positive selection of T cells. With regard to the positive selection of CD8$^+$ T cells, it has also been shown that a single peptide-MHC complex positively selects a diverse and specific CD8 T cell repertoire (Wang et al. 2009). Further evidence about the relationship between positive selection ligands and specificities of post-selection T cells comes from two transgenic mouse lines that each expresses a different single peptide–MHC class I complex (OVA and VSVp). Each mouse line exhibits exclusive TCR repertoires with unique peptide specificities that do not exist in the other mouse line, while the two lines do share some overlapping peptide specificities. From this piece of evidence, we learn that the positive selection in CD8$^+$ T cells is very peptide specific. Each positively selecting ligand may exhibit a different spectrum of positive selection capability, resulting in a unique post-selection repertoire. However, two different positively selecting peptides may both be capable of positively selecting same TCRs. Taken together, positive selection requires the specific recognition of self-peptides to generate a complete T cell repertoire.

A question remains that whether the specific recognition of positively selecting self-peptides may influence the peptide specificities of post-selection T cell repertoire. To answer the question, four Ii-peptide transgenic mouse lines were generated, including Eα, CLIP, CD22, and Rab5a (Barton et al. 2002). The mature CD4[+] T cells in these four Ii-peptide transgenic mouse lines showed different degrees of proliferative responses in mixed lymphocyte cultures. The CLIP- and Rab5a-selected CD4[+] T cells proliferated most strongly, Eα-selected CD4[+] T cells proliferated moderately, and CD22-selected CD4[+] T cells proliferated weakly (Barton et al. 2002). The study convincingly showed that T cells selected by one peptide have different specificities compared with T cells selected by a second peptide (Barton et al. 2002). Similar studies examined the positive selection of T cells that are specific for MCC responses (Liu et al. 1997; Nakano et al. 1997). The CD4[+] T cells were positively selected by MCC peptide, MCC variants, or unrelated Hb peptide (Nakano et al. 1997). The sequence of engineered self-peptides directly influenced the post-selection T cell's capability of responding to the MCC peptide, and TCR usage (Liu et al. 1997; Nakano et al. 1997).

Thus, taken together, the recognition of positively selecting peptides has a certain degree of specificity. Even though a single peptide can induce positive selection of a large population of T cells (one peptide is capable of selecting 10^5 distinct TCR) (Gapin et al. 1998), but no single peptide is capable of selecting a full repertoire (Fig. 1b) (Barton et al. 2002; Barton and Rudensky 1999). When multiple peptides can select a T cell population responsive to the same antigen, the different peptides can select repertoire varying in patterns of fine antigen specificity and TCR usage.

3 The Role of Self-Peptides in Periphery

3.1 Positively Selecting Self-Peptides May Maintain Homeostatic Proliferation and Survival of Peripheral T Cells

In periphery, T cells require continuous low-level TCR interaction with self-peptides to survive (Brocker 1997; Davis et al. 2007; Min and Paul 2005; Morris and Allen 2012). Homeostatic proliferation is driven by a low-affinity interaction with self-peptide/MHC (Sprent et al. 2008; Surh and Sprent 2000, 2008). It remains unknown whether T cell reactivity to self-peptides in periphery relates to the affinity threshold established during positive selection. Positively selecting self-peptides may not only rescue DP thymocytes from programed cell death, but also program the long-term competitive fitness of post-selection T cells in periphery to influence its survival, proliferation, and functional alertness (Ernst et al. 1999; Goldrath and Bevan 1999; Viret et al. 1999) (Fig. 2).

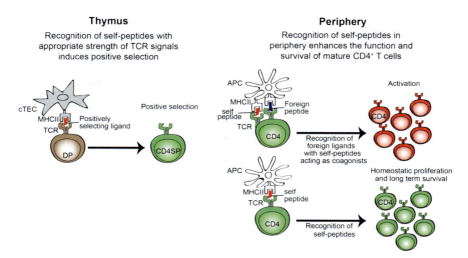

Fig. 2 Positively selecting ligands may promote the positive selection in the thymus, as well as function as coagonists and help maintain mature T cell survival in the periphery

Do positively selecting ligands contribute to naive T cell homeostatic proliferation in periphery? Is the ligand that mediate homeostatic proliferation the same as the positively selecting ligand in thymus? The first evidence came from the homeostatic proliferation experiments by using H-2M-deficient mice (Goldrath and Bevan 1999; Viret et al. 1999). The MHC class II molecules of H-2M-deficient mice are almost exclusively loaded with class II invariant chain peptides (CLIPs) (Fung-Leung et al. 1996; Martin et al. 1996; Miyazaki et al. 1996), so that the CLIP peptide was highly expressed in both thymus, and in periphery. Naive CD4$^+$ cells from B6 mice failed to proliferate in T cell-depleted H-2M-deficient hosts, whereas naive CD4$^+$ cells from H-2M-deficient hosts proliferate strongly (Goldrath and Bevan 1999; Viret et al. 1999). The difference in proliferative responses may result from the possibility that homeostatic proliferation is driven by the positively selecting ligands. H-2M deficient CD4$^+$ T cells were positively selected by high abundant CLIP in the thymus, and therefore were capable of homeostatic proliferation while adoptively transferred to H-2M deficient hosts. On the other hand, wide-type B6 CD4$^+$ T cells were positively selected by a normal self-peptide repertoire. Given the normal self-peptide repertoire is absent in H-2M deficient hosts because the peptide exchange machinery is disrupted, wide-type B6 CD4$^+$ T cells failed to homeostatically proliferate. Additionally, in H-2M-deficient mice, recognition of self-peptide/MHC may maintain the survival and repertoire of mature CD4SP T cells (Kieper et al. 2004; Moses et al. 2003). Similar evidence comes from studies of CD8$^+$ T cells using TAP-deficient mice. OVA-specific CD8 T cells failed to proliferate when transferred into TAP-deficient mice, but they would proliferate when transferred into transgenic mice expressing positively selecting altered peptide transgenes of OVA (Goldrath and Bevan 1999).

More recently, using the naturally occurring self-peptides for CD4$^+$ T cell selection (gp250 peptide for positively selecting AND TCR and Gag-Pol peptide for 5C.C7 TCR) (Lo et al. 2009; Ebert et al. 2009), the relationship of the ligand that positively select and that to mediate homeostatic proliferation was tested. CFSE-labeled naive AND CD4$^+$ T cells were adoptively transferred to chronic lymphopenic B6.$Rag1^{-/-}$H-2^k recipients, and mice were injected intraperitoneally with additional positively selecting ligand gp250 or non-selecting control peptide Hb. Two- to ninefold more AND CD4$^+$ T cells were recovered from the gp250-injected mice (Lo et al. 2009). A single mutation of the gp250 TCR contact residues disrupt gp250 ability to enhance the survival of AND CD4$^+$ T cells, suggesting the enhanced survival is gp250-peptide specific. Interestingly, the recognition of self-peptides in periphery has the identical high degree of specificity to that observed for positive selection (Lo et al. 2009). In the case of positively selecting self-peptide Gag-Pol and 5C.C7 T cells, Gag-Pol peptide can also drive the homeostatic proliferation of 5C.C7 CD4$^+$ T cells (Singh et al. 2012; Walker 2012), supporting the hypothesis that the crucial role of positively selecting ligands extends beyond the thymus into periphery.

3.2 Positively Selecting Self-Peptides May Function as Co-agonists to Augment Functional Sensitivity of Peripheral T Cells

In addition to homeostatic proliferation and maintenance of peripheral T cells, positively selecting ligands can also directly contribute to the functional alertness and responsiveness of naive T cells to cognate antigens. The first notion of such a possibility was suggested by the observation that self-peptides were co-localized with agonist peptides at the immunological synapse (Wulfing et al. 2002). The co-localization of self-peptides with agonist peptides was shown to participate in T cell activation by acting as a co-agonist in both MHC class I and class II systems (Irvine et al. 2002; Krogsgaard et al. 2005). The pseudodimer model suggests CD4 core-ceptor allows certain self-peptide/MHC complexes to contribute to T cell activation, thus functioning as a coagonist. With 5C.C7 T cells (Irvine et al. 2002; Krogsgaard et al. 2005), three out of seven self-peptides were demonstrated to act as co-agonists and enhance T cell activation when presented in conjunction with small amounts of agonist peptide. Despite the unlikelihood of the CD8 co-receptor forming a similar pseudodimer, self-peptides have been shown to also function as coagonists in class I-restricted responses (Yachi et al. 2005). These observations raise the possibility that positively selecting peptides encountered in the thymus may continue to have profound effects on T cell responses in the periphery. In the gp250/AND system, the positively selecting self-peptide gp250 can enhance naive AND T cell activation to lower concentration of agonist MCC (Lo et al. 2009). Similarly, in the case of 5C.C7, the positively selecting ligand Gag-Pol can enhance naive 5C.C7 T cell activation in response to agonist MCC peptide, while other nonrelated peptides lacked such capability (Juang et al. 2010). The validity of the co-agonist model is still debated, and resolution of this issue awaits a determination of the precise number of TCRs (one or more than one) in the initial T cell recognition with limiting number (physiological numbers) of peptide/MHC complexes.

3.3 Positive Selection Signals May Program the Long-Term Survival and Function of Peripheral T Cells

So how does the positive selection "experience" become a critical determinant to influence peripheral T cell function and survival? Two studies in $CD8^+$ T cells have related positive selecting signaling events to T cell maintenance and functional capability in periphery by using $CD8^+$ TCR transgenic mouse lines which have a different affinity for self-peptide/MHC complexes (Cho et al. 2010; Sinclair et al. 2011), with OT-I $CD8^+$ T cells having the highest sensitivity, followed by 2C $CD8^+$ T cells, where HY $CD8^+$ T cells have very low or undetectable sensitivity to homeostatic proliferation. The $CD8^+$ T cell homeostatic proliferation relies on TCR and IL-7 signaling. The propensity for T cells to undergo homeostatic proliferation correlates with their intrinsic TCR affinity for self MHC ligands. Thus, for $CD8^+$ T cells, naive cells from OT-1 and 2C TCR transgenic mouse lines have relatively high affinity for self-peptide/MHC ligands, while the HY $CD8^+$ T cells have the relatively low self reactivity, such that they fail to homeostatically proliferate. The strength of positively selecting signals dynamically regulate IL-7Rα abundance and CD5 surface expression (Sinclair et al. 2011). IL-7 receptor signals are critical for all peripheral T cell subsets (Sinclair et al. 2011), and CD5 constitutively associates with SHP-1 to negative regulate TCR signals (Azzam et al. 1998). The stronger the positive selection signals, the higher the expression of IL-7R and CD5 express on the cell surface on mature T cells (Cho et al. 2010; Sinclair et al. 2011). The variation in IL-7R and CD5 expression may determine the capability of homeostatic survival and proliferation of these three TCR transgenic $CD8^+$ T cell clones. Therefore, positive selection programs IL-7Rα and CD5 expression on mature T cells to influence the ability of new generated T cells to be maintained within the peripheral T cell repertoire (Fig. 3).

TCR affinity for self-peptide-MHC complexes not only influences their potential for homeostatic proliferation, but also tunes T cell activation threshold in response to cognate antigen. Our laboratory generated two TCR transgenic mice, LLO56 and LLO118, specific for an immunodominant Listeria epitope (listeriolysin 190–205) and only differing by 15 amino acids in their TCR sequences (Weber et al. 2012). These cells differed only in their CD5 levels, and showed dramatically different in vivo responses against Listeria infection. LLO56, with higher CD5 expression, has a significantly stronger recall response, whereas LLO118, with lower CD5 surface expression, mediated a better primary response.

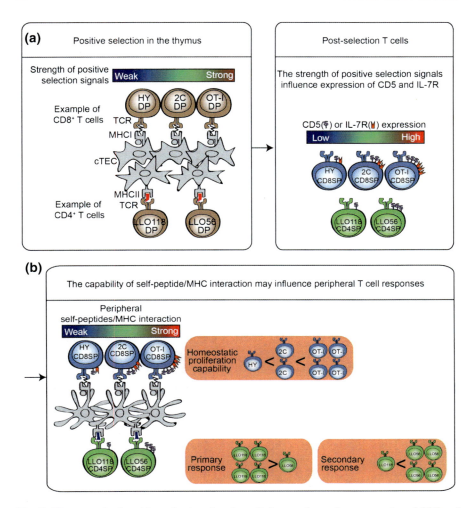

Fig. 3 The strength of positive selection signal may influence the surface expression of CD5 and IL-7Rα on the post-selection mature T cells, and therefore affect peripheral T cell function, such as homeostatic proliferation and cell responses against pathogens

More interestingly, while sorting out LLO118 cells that express similar CD5 levels as LLO56 cells, CD5^{hi} LLO118 cells became good recall responders as well (Fig. 3). This observation directly correlates CD5 expression level and peripheral function. While positive selection signals program CD5 surface expression, the competitive fitness and peripheral function are also imprinted in mature T cells.

4 Conclusion

Accumulated evidence strongly suggests that the recognition of a positively selecting ligand is not promiscuous, or simply a passive rescue process to prevent cell death by neglect or by negative selection. Instead, individual positively selecting ligands may enrich unique repertoires of mature T cells. One positively selecting self-peptide has to select more than several thousand T cells, and the TCR recognition of positively selecting self-peptides has a high degree of specificity. The specific interaction between individual TCRs and self-peptide/MHC complexes may determine the expression levels of signaling molecules and cytokine receptors, such as CD5 and IL-7Rα, therefore affecting functional sensitivity and competitive fitness of peripheral T cells. Similarly, one positively selecting ligand may differ with the other ligands in terms of the capability of positive selection, that some self-peptides may be better selectors to positively select a broader TCR repertoire, and some might select a narrower repertoire.

Acknowledgments We thank G. Morris and D. Donermeyer for their critical reading of the manuscript and comments.

References

Alarcon B, and van Santen HM (2010) Two receptors, two kinases, and T cell lineage determination. Sci Signal 3:pe11

Azzam HS, Grinberg A, Lui K, Shen H, Shores EW, Love PE (1998) CD5 expression is developmentally regulated by T cell receptor (TCR) signals and TCR avidity. J Exp Med 188:2301–2311

Barton GM, Rudensky AY (1999) Requirement for diverse, low-abundance peptides in positive selection of T cells. Science 283:67–70

Barton GM, Beers C, deRoos P, Eastman SR, Gomez ME, Forbush KA, Rudensky AY (2002) Positive selection of self-MHC-reactive T cells by individual peptide-MHC class II complexes. Proc Natl Acad Sci U S A 99:6937–6942

Bluestone JA, Pardoll D, Sharrow SO, Fowlkes BJ (1987) Characterization of murine thymocytes with CD3-associated T-cell receptor structures. Nature 326:82–84

Brocker T (1997) Survival of mature CD4 T lymphocytes is dependent on major histocompatibility complex class II-expressing dendritic cells. J Exp Med 186:1223–1232

Casrouge A, Beaudoing E, Dalle S, Pannetier C, Kanellopoulos J, Kourilsky P (2000) Size estimate of the alpha beta TCR repertoire of naive mouse splenocytes. J Immunol 164:5782–5787

Chiang YJ, Kole HK, Brown K, Naramura M, Fukuhara S, Hu RJ, Jang IK, Gutkind JS, Shevach E, Gu H (2000) Cbl-b regulates the CD28 dependence of T-cell activation. Nature 403:216–220

Chmielowski B, Muranski P, Kisielow P, Ignatowicz L (2000) On the role of high- and low-abundance class II MHC-peptide complexes in the thymic positive selection of CD4(+) T cells. Int Immunol 12:67–72

Cho JH, Kim HO, Surh CD, Sprent J (2010) T cell receptor-dependent regulation of lipid rafts controls naive CD8+ T cell homeostasis. Immunity 32:214–226

Davey GM, Schober SL, Endrizzi BT, Dutcher AK, Jameson SC, Hogquist KA (1998) Preselection thymocytes are more sensitive to T cell receptor stimulation than mature T cells. J Exp Med 188:1867–1874

Davis MM, Krogsgaard M, Huse M, Huppa J, Lillemeier BF, Li QJ (2007) T cells as a self-referential, sensory organ. Annu Rev Immunol 25:681–695

Ebert PJ, Jiang S, Xie J, Li QJ, Davis MM (2009) An endogenous positively selecting peptide enhances mature T cell responses and becomes an autoantigen in the absence of microRNA miR-181a. Nat Immunol 10:1162–1169

Ebert PJ, Li QJ, Huppa JB, Davis MM (2010) Functional development of the T cell receptor for antigen. Prog Mol Biol Transl Sci 92:65–100

Eck SC, Zhu P, Pepper M, Bensinger SJ, Freedman BD, Laufer TM (2006) Developmental alterations in thymocyte sensitivity are actively regulated by MHC class II expression in the thymic medulla. J Immunol 176:2229–2237

Ernst B, Lee DS, Chang JM, Sprent J, Surh CD (1999) The peptide ligands mediating positive selection in the thymus control T cell survival and homeostatic proliferation in the periphery. Immunity 11:173–181

Fortier MH, Caron E, Hardy MP, Voisin G, Lemieux S, Perreault C, Thibault P (2008) The MHC class I peptide repertoire is molded by the transcriptome. J Exp Med 205:595–610

Fukui Y, Ishimoto T, Utsuyama M, Gyotoku T, Koga T, Nakao K, Hirokawa K, Katsuki M, Sasazuki T (1997) Positive and negative CD4+ thymocyte selection by a single MHC class II/peptide ligand affected by its expression level in the thymus. Immunity 6:401–410

Fukui Y, Hashimoto O, Inayoshi A, Gyotoku T, Sano T, Koga T, Gushima T, Sasazuki T (1998) Highly restricted T cell repertoire shaped by a single major histocompatibility complex-peptide ligand in the presence of a single rearranged T cell receptor beta chain. J Exp Med 188:897–907

Fung-Leung WP, Surh CD, Liljedahl M, Pang J, Leturcq D, Peterson PA, Webb SR, Karlsson L (1996) Antigen presentation and T cell development in H2-M-deficient mice. Science 271:1278–1281

Gapin L, Fukui Y, Kanellopoulos J, Sano T, Casrouge A, Malier V, Beaudoing E, Gautheret D, Claverie JM, Sasazuki T et al (1998) Quantitative analysis of the T cell repertoire selected by a single peptide-major histocompatibility complex. J Exp Med 187:1871–1883

Goldrath AW, Bevan MJ (1999) Selecting and maintaining a diverse T-cell repertoire. Nature 402:255–262

Grubin CE, Kovats S, deRoos P, Rudensky AY (1997) Deficient positive selection of CD4 T cells in mice displaying altered repertoires of MHC class II-bound self-peptides. Immunity 7:197–208

Havran WL, Poenie M, Kimura J, Tsien R, Weiss A, Allison JP (1987) Expression and function of the CD3-antigen receptor on murine CD4+8+ thymocytes. Nature 330:170–173

Hernandez-Hoyos G, Sohn SJ, Rothenberg EV, Alberola-Ila J (2000) Lck activity controls CD4/CD8 T cell lineage commitment. Immunity 12:313–322

Hogquist KA, Tomlinson AJ, Kieper WC, McGargill MA, Hart MC, Naylor S, Jameson SC (1997) Identification of a naturally occurring ligand for thymic positive selection. Immunity 6:389–399

Honey K, Rudensky AY (2003) Lysosomal cysteine proteases regulate antigen presentation. Nat Rev Immunol 3:472–482

Honey K, Nakagawa T, Peters C, Rudensky A (2002) Cathepsin L regulates CD4+T cell selection independently of its effect on invariant chain: a role in the generation of positively selecting peptide ligands. J Exp Med 195:1349–1358

Hsieh CS, deRoos P, Honey K, Beers C, Rudensky AY (2002) A role for cathepsin L and cathepsin S in peptide generation for MHC class II presentation. J Immunol 168:2618–2625

Ignatowicz L, Kappler J, Marrack P (1996) The repertoire of T cells shaped by a single MHC/peptide ligand. Cell 84:521–529

Ignatowicz L, Rees W, Pacholczyk R, Ignatowicz H, Kushnir E, Kappler J, Marrack P (1997) T cells can be activated by peptides that are unrelated in sequence to their selecting peptide. Immunity 7:179–186

Irvine DJ, Purbhoo MA, Krogsgaard M, Davis MM (2002) Direct observation of ligand recognition by T cells. Nature 419:845–849

Jameson SC, Hogquist KA, Bevan MJ (1995) Positive selection of thymocytes. Annu Rev Immunol 13:93–126

Juang J, Ebert PJ, Feng D, Garcia KC, Krogsgaard M, Davis MM (2010) Peptide-MHC heterodimers show that thymic positive selection requires a more restricted set of self-peptides than negative selection. J Exp Med 207:1223–1234

Kieper WC, Burghardt JT, Surh CD (2004) A role for TCR affinity in regulating naive T cell homeostasis. J Immunol 172:40–44

Krogsgaard M, Li QJ, Sumen C, Huppa JB, Huse M, Davis MM (2005) Agonist/endogenous peptide-MHC heterodimers drive T cell activation and sensitivity. Nature 434:238–243

Legname G, Seddon B, Lovatt M, Tomlinson P, Sarner N, Tolaini M, Williams K, Norton T, Kioussis D, Zamoyska R (2000) Inducible expression of a p56Lck transgene reveals a central role for Lck in the differentiation of CD4 SP thymocytes. Immunity 12:537–546

Li QJ, Chau J, Ebert PJ, Sylvester G, Min H, Liu G, Braich R, Manoharan M, Soutschek J, Skare P et al (2007) miR-181a is an intrinsic modulator of T cell sensitivity and selection. Cell 129:147–161

Liu CP, Parker D, Kappler J, Marrack P (1997) Selection of antigen-specific T cells by a single IEk peptide combination. J Exp Med 186:1441–1450

Lo WL, Felix NJ, Walters JJ, Rohrs H, Gross ML, Allen PM (2009) An endogenous peptide positively selects and augments the activation and survival of peripheral CD4+T cells. Nat Immunol 10:1155–1161

Lo WL, Donermeyer DL, Allen PM (2012) A voltage-gated sodium channel is essential for the positive selection of CD4(+) T cells. Nat Immunol 13:880–887

Lorenz RG, Allen PM (1988) Direct evidence for functional self-protein/Ia-molecule complexes in vivo. Proc Natl Acad Sci U S A 85:5220–5223

Malherbe L, Hausl C, Teyton L, McHeyzer-Williams MG (2004) Clonal selection of helper T cells is determined by an affinity threshold with no further skewing of TCR binding properties. Immunity 21:669–679

Mariathasan S, Zakarian A, Bouchard D, Michie AM, Zuniga-Pflucker JC, Ohashi PS (2001) Duration and strength of extracellular signal-regulated kinase signals are altered during positive versus negative thymocyte selection. J Immunol 167:4966–4973

Marrack P, Ignatowicz L, Kappler JW, Boymel J, Freed JH (1993) Comparison of peptides bound to spleen and thymus class II. J Exp Med 178:2173–2183

Martin WD, Hicks GG, Mendiratta SK, Leva HI, Ruley HE, Van Kaer L (1996) H2-M mutant mice are defective in the peptide loading of class II molecules, antigen presentation, and T cell repertoire selection. Cell 84:543–550

McNeil LK, Starr TK, Hogquist KA (2005) A requirement for sustained ERK signaling during thymocyte positive selection in vivo. Proc Natl Acad Sci U S A 102:13574–13579

McNeill L, Salmond RJ, Cooper JC, Carret CK, Cassady-Cain RL, Roche-Molina M, Tandon P, Holmes N, Alexander DR (2007) The differential regulation of Lck kinase phosphorylation sites by CD45 is critical for T cell receptor signaling responses. Immunity 27:425–437

Min B, Paul WE (2005) Endogenous proliferation: burst-like CD4 T cell proliferation in lymphopenic settings. Semin Immunol 17:201–207

Miyazaki T, Wolf P, Tourne S, Waltzinger C, Dierich A, Barois N, Ploegh H, Benoist C, Mathis D (1996) Mice lacking H2-M complexes, enigmatic elements of the MHC class II peptide-loading pathway. Cell 84:531–541

Moran AE, Hogquist KA (2012) T-cell receptor affinity in thymic development. Immunology 135:261–267

Morris GP, Allen PM (2012) How the TCR balances sensitivity and specificity for the recognition of self and pathogens. Nat Immunol 13:121–128

Moses CT, Thorstenson KM, Jameson SC, Khoruts A (2003) Competition for self ligands restrains homeostatic proliferation of naive CD4 T cells. Proc Natl Acad Sci U S A 100:1185–1190

Murata S, Sasaki K, Kishimoto T, Niwa S, Hayashi H, Takahama Y, Tanaka K (2007) Regulation of CD8 + T cell development by thymus-specific proteasomes. Science 316:1349–1353

Nakagawa T, Roth W, Wong P, Nelson A, Farr A, Deussing J, Villadangos JA, Ploegh H, Peters C, Rudensky AY (1998) Cathepsin L: critical role in Ii degradation and CD4 T cell selection in the thymus. Science 280:450–453

Nakano N, Rooke R, Benoist C, Mathis D (1997) Positive selection of T cells induced by viral delivery of neopeptides to the thymus. Science 275:678–683

Naramura M, Kole HK, Hu RJ, Gu H (1998) Altered thymic positive selection and intracellular signals in Cbl-deficient mice. Proc Natl Acad Sci U S A 95:15547–15552

Nitta T, Murata S, Sasaki K, Fujii H, Ripen AM, Ishimaru N, Koyasu S, Tanaka K, Takahama Y (2010) Thymoproteasome shapes immunocompetent repertoire of CD8+T cells. Immunity 32:29–40

Plas DR, Williams CB, Kersh GJ, White LS, White JM, Paust S, Ulyanova T, Allen PM, Thomas ML (1999) Cutting edge: the tyrosine phosphatase SHP-1 regulates thymocyte positive selection. J Immunol 162:5680–5684

Rothenberg EV, Moore JE, Yui MA (2008) Launching the T-cell-lineage developmental programme. Nat Rev Immunol 8:9–21

Saini M, Sinclair C, Marshall D, Tolaini M, Sakaguchi S, Seddon B (2010) Regulation of Zap70 expression during thymocyte development enables temporal separation of CD4 and CD8 repertoire selection at different signaling thresholds. Sci Signal 3:ra23

Santori FR, Kieper WC, Brown SM, Lu Y, Neubert TA, Johnson KL, Naylor S, Vukmanovic S, Hogquist KA, Jameson SC (2002) Rare, structurally homologous self-peptides promote thymocyte positive selection. Immunity 17:131–142

Schmedt C, Saijo K, Niidome T, Kuhn R, Aizawa S, Tarakhovsky A (1998) Csk controls antigen receptor-mediated development and selection of T-lineage cells. Nature 394:901–904

Sebzda E, Kundig TM, Thomson CT, Aoki K, Mak SY, Mayer JP, Zamborelli T, Nathenson SG, Ohashi PS (1996) Mature T cell reactivity altered by peptide agonist that induces positive selection. J Exp Med 183:1093–1104

Sinclair C, Saini M, van der Loeff IS, Sakaguchi S, Seddon B (2011) The long-term survival potential of mature T lymphocytes is programmed during development in the thymus. Sci Signal 4:ra77

Singer A, Adoro S, Park JH (2008) Lineage fate and intense debate: myths, models and mechanisms of CD4- versus CD8-lineage choice. Nat Rev Immunol 8:788–801

Singh NJ, Bando JK, Schwartz RH (2012) Subsets of nonclonal neighboring CD4+T cells specifically regulate the frequency of individual antigen-reactive T cells. Immunity 37:735–746

Sprent J, Cho JH, Boyman O, Surh CD (2008) T cell homeostasis. Immunol Cell Biol 86:312–319

Stephen TL, Tikhonova A, Riberdy JM, Laufer TM (2009) The activation threshold of CD4+T cells is defined by TCR/peptide-MHC class II interactions in the thymic medulla. J Immunol 183:5554–5562

Stritesky GL, Jameson SC, Hogquist KA (2012) Selection of self-reactive T cells in the thymus. Annu Rev Immunol 30:95–114

Surh CD, Sprent J (2000) Homeostatic T cell proliferation: how far can T cells be activated to self-ligands? J Exp Med 192:F9–F14

Surh CD, Sprent J (2008) Homeostasis of naive and memory T cells. Immunity 29:848–862

Surh CD, Lee DS, Fung-Leung WP, Karlsson L, Sprent J (1997) Thymic selection by a single MHC/peptide ligand produces a semidiverse repertoire of CD4+T cells. Immunity 7:209–219

Suri A, Vidavsky I, van der Drift K, Kanagawa O, Gross ML, Unanue ER (2002) In APCs, the autologous peptides selected by the diabetogenic I-Ag7 molecule are unique and determined by the amino acid changes in the P9 pocket. J Immunol 168:1235–1243

Takahama Y, Nitta T, Mat Ripen A, Nitta S, Murata S, Tanaka K (2010) Role of thymic cortex-specific self-peptides in positive selection of T cells. Semin Immunol 22:287–293

Viret C, Wong FS, Janeway CA Jr (1999) Designing and maintaining the mature TCR repertoire: the continuum of self-peptide:self-MHC complex recognition. Immunity 10:559–568

Walker LS (2012) Maintaining a competitive edge: new rules for peripheral T cell homeostasis. Immunity 37:598–600

Wang L, Bosselut R (2009) CD4-CD8 lineage differentiation: Thpok-ing into the nucleus. J Immunol 183:2903–2910

Wang B, Primeau TM, Myers N, Rohrs HW, Gross ML, Lybarger L, Hansen TH, Connolly JM (2009) A single peptide-MHC complex positively selects a diverse and specific CD8 T cell repertoire. Science 326:871–874

Wang D, Zheng M, Lei L, Ji J, Yao Y, Qiu Y, Ma L, Lou J, Ouyang C, Zhang X et al (2012) Tespa1 is involved in late thymocyte development through the regulation of TCR-mediated signaling. Nat Immunol 13:560–568

Weber KS, Li QJ, Persaud SP, Campbell JD, Davis MM, Allen PM (2012) Distinct CD4+helper T cells involved in primary and secondary responses to infection. Proc Natl Acad Sci U S A 109:9511–9516

Werlen G, Hausmann B, Palmer E (2000) A motif in the alphabeta T-cell receptor controls positive selection by modulating ERK activity. Nature 406:422–426

Wiest DL, Yuan L, Jefferson J, Benveniste P, Tsokos M, Klausner RD, Glimcher LH, Samelson LE, Singer A (1993) Regulation of T cell receptor expression in immature CD4+CD8+ thymocytes by p56lck tyrosine kinase: basis for differential signaling by CD4 and CD8 in immature thymocytes expressing both coreceptor molecules. J Exp Med 178:1701–1712

Wulfing C, Sumen C, Sjaastad MD, Wu LC, Dustin ML, Davis MM (2002) Costimulation and endogenous MHC ligands contribute to T cell recognition. Nat Immunol 3:42–47

Yachi PP, Ampudia J, Gascoigne NR, Zal T (2005) Nonstimulatory peptides contribute to antigen-induced CD8-T cell receptor interaction at the immunological synapse. Nat Immunol 6:785–792

Central Tolerance Induction

Maria L. Mouchess and Mark Anderson

Abstract A critical function of the thymus is to help enforce tolerance to self. The importance of central tolerance in preventing autoimmunity has been enlightened by a deeper understanding of the interactions of developing T cells with a diverse population of thymic antigen presenting cell populations. Furthermore, there has been rapid progress in our understanding of how autoreactive T cell specificities are diverted into the T regulatory lineage. Here we review and highlight the recent progress in how tolerance is imposed on the developing thymocyte repertoire.

Contents

1 Central Tolerance Induction	69
1.1 Central Tolerance Through Deletion	70
1.2 Tolerance Through Treg Selection	76
2 Concluding Remarks	80
References	80

1 Central Tolerance Induction

Immune tolerance is an essential process in the immune system to prevent untoward responses to self. The thymus not only provides the proper selecting environment for the positive selection of T cells, but also plays a critical role in promoting tolerance of the developing T cell repertoire to self. Central tolerance

---

M. L. Mouchess · M. Anderson (✉)
Diabetes Center, University of California-San Francisco, Box 0540 San Francisco, CA 94143, USA
e-mail: manderson@diabetes.ucsf.edu

Current Topics in Microbiology and Immunology (2014) 373: 69–86
DOI: 10.1007/82_2013_321
© Springer-Verlag Berlin Heidelberg 2013
Published Online: 10 May 2013

plays an integral role in immune tolerance along with a net of other peripheral tolerance mechanisms that together maintain an immune repertoire that is fit for response to a diverse array of potential foreign antigens but is unable to respond to self-antigens.

It has now become widely appreciated that during thymic selection, conventional $\alpha\beta$ chain-expressing T cells with significant autoreactivity are tolerized mainly by deletion or induction into the Foxp3+ T regulatory lineage. The rules and mechanisms that lead to one fate over another are still being elucidated, however, there has been recent rapid progress in this area. Furthermore, the selecting environment present in the thymic medulla has also been an area of recent intensive investigation. Here in this review, we highlight some of the recent progress in our understanding of how central tolerance is imposed on the conventional $\alpha\beta$ T cell repertoire.

1.1 Central Tolerance Through Deletion

Developing thymocytes are exposed to a wide array of self-antigens within the thymus and those T cells that can bind to self-antigen peptide/MHC complexes with high affinity are removed from the immune repertoire through deletional mechanisms. Thus, there are a number of important factors involved for this process to play out. First, is the timing and expression of a functional TCR during development, second is the timing and expression of self-antigen peptide/MHC ligands by APCs present in the thymus, and finally are the factors and pathways that allow for a deletional/apoptotic death to occur in autoreactive thymocytes.

As outlined by Allen and colleagues elsewhere in this book, developing thymocytes go through a coordinated series of developmental steps that involve the generation of a unique $\alpha\beta$ chain TCR complex. Pre-T cells are recruited into the thymus and enter the thymic cortex and rearrangement of the TCR β chain occurs through RAG-mediated recombination (Schatz et al. 1989; Oettinger et al. 1990). If a functional β chain is formed, it complexes with a pre-T α chain, migrates to the cell surface, and this helps instruct the rearrangement of the TCR α chain (Fehling et al. 1995). If such an α chain can complex with the rearranged TCR β chain, a potentially functional TCR is then expressed at the cell surface. At this stage of development such T cells co-express both CD4 and CD8 and are termed Double Positives (DPs). Such DPs then proceed through an important step of selection termed positive selection where individual clones are exposed to self-peptide/MHC complexes present in the cortex. Individual T cell clones that have relatively low affinity for these complexes are allowed to survive, whereas those clones with no affinity for self-peptide/MHC die by neglect at this selection step. T cells surviving beyond this stage develop into CD4 or CD8 Single Positive cells (SPs) that then migrate into the thymic medulla, where they are further selected and those clones with high affinity for self-peptide/MHC complexes present on APCs present in this compartment are deleted or diverted into the Treg lineage (discussed later).

Although it had long been postulated that deletion of autoreactive clones was a major mechanism of central tolerance (Burnet 1958), clonal deletion was only first clearly demonstrated in vivo by examining superantigen reactive T cells (Kappler et al. 1987). Part of the reason for this was the lack of sophisticated tools to detect deletion of autoreactive thymocytes in the polyclonal repertoire. The development of T cell receptor transgenic mice in the late 1980s led to the development of experimental tools that allowed a more refined assessment of thymic deletion for single individual TCR clones. For example, TCR transgenic mice with specificity for the male HY self-antigen again confirmed the existence of a thymic deletional mechanism when thymocytes were exposed to their cognate self-antigen (Kisielow et al. 1988). Furthermore, injection of antigens that individual TCR transgenic were specific for demonstrated again the existence of thymocyte deletion when antigens were present in the thymus (Murphy et al. 1990; Fowlkes et al. 1988).

The mechanisms that lead to thymocyte death after antigen encounter are generally thought to involve controlled apoptotic mechanisms (Sohn et al. 2007). The orphan nuclear hormone receptor, Nur77 appears to be a key regulator of such apoptotic death. Nur77 is upregulated in thymocytes when they are exposed to high affinity ligands (Calnan et al. 1995; Cho et al. 2003) and although not completely worked out, it appears to exert an effect on pro-apoptotic mitochondrial proteins (Thompson and Winoto 2008; Fassett et al. 2012). In addition, to Nur77, the pro-apoptotic BH3-only protein Bim has been linked to thymic deletion (Bouillet et al. 2002) in that Bim-deficient T cells have shown to be resistant to thymic deletion in a number of models. Recently, a more thorough screen of other BH3-only proteins for their role in thymic deletion was explored and the BH3-only protein PUMA was also found to contribute to this process (Gray et al. 2012). Interestingly, Bim/PUMA double deficient mice demonstrate a more robust defect in thymic deletion than single deficient mice of either genotype and thus suggest that these molecules work in a complementary fashion to drive deletion.

As outlined above, thymocytes require low affinity TCR signals to be positively selected, yet high affinity TCR signals appear to drive deletion. Thus, an interesting question remains as to the differences in the TCR signaling cascade that promote positive versus negative selection. Work by Palmer and colleagues with a panel of Ovalbumin peptide mimotopes with a range of affinities for the OT-I TCR has demonstrated an exquisite line of TCR affinity for peptide/MHC that demarcates positive selection from negative selection (Daniels et al. 2006). Downstream of this affinity, there appears to be a number of differential signaling cascades that have been implicated in the distinction between positive and negative selection. For example, it has been suggested that high affinity TCR ligation leads to a conformational change in the tail of the CD3 epsilon chain that allows for the adapter molecule Lck to bind and interact with the chain (Gil et al. 2005), however, there remains conflicting data on this model (Nika et al. 2010). Differential activation of the proximal TCR signaling molecules JNK and ERK have also been implicated in positive versus negative selection signals (McNeil et al. 2005). In addition, a recent report demonstrated differential intracellular localization of Ras and MAP-kinase signaling components during positive versus negative selection signaling (Daniels et al. 2006), again reinforcing differences in TCR signaling cascades.

Thymic APC Subsets and tolerance. There are multiple Antigen Presenting Cell (APC) types that the developing thymocyte repertoire interacts with during maturation and each of these cell types likely play a role in imparting central tolerance. These major APC cell types included Cortical Thymic Epithelial Cells (cTECs), Medullary Epithelial Cells (mTECs), and Dendritic Cells (DCs) (see Fig. 1). All three of these major subsets display specialized expression of MHC Class II, antigen processing and presentation machinery, and costimulatory ligand expression (i.e., CD80 and/or CD86). Each subset appears to have sub-specialized properties with cTECs being a critical regulator of positive selection and mTECs and DCs being critical regulators of negative selection.

After pre-T cells enter the thymus at the Cortico Medullary Junction (CMJ), they migrate into the cortex where they interact with cTECs. cTECs are epithelial cells derived from the endoderm of the third brachial pouch that express both MHC Class II and Class I molecules complexed with a broad array of self-peptides and interaction of T cells with these ligands is critical for positive selection (Starr et al. 2003). cTECs are also capable of mediating negative selection, however, the timing of when high levels of TCR expression are occurring during T cell development in this compartment also play into whether cTECs drive negative selection. Early work with the HY-TCR transgenic demonstrated that HY-specific T cells were likely being deleted by HY expressing cTECs, however, subsequent studies revealed that this deletion may have been an artifact of early expression of the TCR due to the transgene utilized (Baldwin et al. 2005). Nonetheless, for certain ubiquitously expressed antigens like HY or superantigen, it remains likely that cTECs can delete T cells with specificity for these antigens. Recently, a specialized property of cTECs was demonstrated showing that they express a specialized subunit of the immunoproteasome called $\beta 5t$ (Murata et al. 2007). The proteasome helps load MHC class I molecules with peptides and the existence of specialized subunit uniquely expressed in cTECs suggests that a unique array of MHC class I associated peptides are generated in cTECs that are not present elsewhere. In addition, the presence of $\beta 5t$ appears to generate an array of peptides that are less stable in the MHC class I binding groove, which may be ideal for low affinity MHC/peptide complexes for positive selection. In support of this notion, $\beta 5t$-deficient mice display a defect in the positive selection of $CD8^+$ T cells. In addition, the lack of peripheral expression of $\beta 5t$ may be part of a fail-safe mechanism for maintaining tolerance in that there is no way to generate the identical MHC/peptide complex in the immunological periphery which could provoke an autoimmune response.

After thymocytes receive positively selecting signals, they then migrate into the thymic medulla, in part, by upregulating the chemokine receptor CCR7 (Kwan and Killeen 2004; Ueno et al. 2004; Yin et al. 2007), where they further interact with the mTEC and DC APC populations that are present in this compartment. Interestingly, there has been recent rapid progress in demonstrating the capacity of mTECs to have the unique property to express a broad array of Tissue-Specific

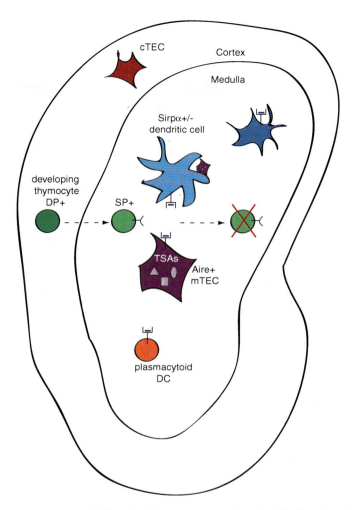

Fig. 1 Antigen presenting cells involved in negative selection of $\alpha\beta$ Tcells in the thymus. The two major types of antigen presenting cells that mediate deletion of autoreactive T cells include dendritic cells (DCs) and medullary thymic epithelial cells (mTECs). High affinity interactions between MHC-antigen complexes with T cell receptors (TCRs) on single positive T cells lead to deletion. Among the thymic DC populations, Sirpa+ DCs and plasmacytoid DCs (pDCs) have been shown to have migratory capacities and are likely to be sources of peripheral antigen. A subset of mTECs express the transcriptional regulator Aire which allows them to express an array of tissue-specific antigens (TSAs). Conventional DCs and mTECs express MHCI and MHCII molecules on their surface and can present TSAs, as cDCs have been shown to present TSAs through antigen transfer from mTECs. Lastly, cortical epithelial cells (cTECs) are critical for positive selection of thymocytes

self-Antigens (TSAs) that is in part under the control of the Autoimmune Regulator (Aire) (Metzger and Anderson 2011). During the 1990s, Hanahan and colleagues have suggested that there were relatively rare thymic medullary cells that had the capacity to express the pancreatic islet-specific protein, insulin at detectable levels (Smith et al. 1997). Subsequent to this, Derbinski and Kyewski demonstrated that mTECs had the unique capacity to express a broad array of TSAs when compared to other thymic APC populations (Derbinski et al. 2001). Part of the molecular answer of how such TSA expression could occur in mTECs came with the discovery that the Aire protein was highly expressed in mTECs and that TSA expression was in part, dependent on Aire (Anderson et al. 2002). *AIRE* was originally identified as the defective gene in patients with the rare autosomal recessive Mendelian syndrome called APS1 (for Autoimmune Polyglandular Syndrome Type 1) (Aaltonen 1997). APS1 patients develop an array of organ-specific autoimmune diseases that frequently target endocrine organs. Studies on the mouse model of APS1, have demonstrated that Aire-deficient mTECs fail to express many TSAs and a direct link between this and a failure in deletion of an autoreactive T cell specific for this type of TSA has now been established (Su et al. 2008; Shum et al. 2009; DeVoss et al. 2010; Taniguchi et al. 2012). Thus, it appears that TSA expression by mTECs is an essential component of tolerance as a defect in properly driving this process leads to an organ-specific autoimmune process.

The mTEC compartment has also been shown to be phenotypically diverse and dynamic. mTECs can be broadly segregated into MHC class $II^{lo}/CD80^{lo}$ (mTEC lo) and MHC class $II^{hi}/CD80^{hi}$ (mTEC hi) subsets (Derbinski et al. 2005). Interestingly, AIRE-expression is restricted to the mTEC hi subset and mTEC hi cells demonstrate a broader array of TSA expression than mTEC lo cells (Derbinski et al. 2005). In BrdU labeling experiments, mTECs appear to have a turnover rate of two weeks and mTEC lo cells have been shown to give rise to mTEC hi cells in fetal thymic organ culture experiments (Gray et al. 2007; Rossi et al. 2007). Thus, it appears that mTECs are in a state of constant turnover and replacement. On an individual cell basis, it appears that TSA expression is stochastic in that there is heterogeneity between individual cells in which TSAs are expressed (Derbinski et al. 2008; Villasenor et al. 2008) and this may help ensure that there is a broad array of self-antigen present across the thymic medulla at all times. It also appears that thymocytes reside in the medulla for an extended period of time with recent estimates being around a four-day transit time (McCaughtry et al. 2007). Thus, thymocytes in the medulla may go through extensive scanning of mTECs and DCs to also help ensure exposure to self-antigens.

The maturation and development of mTECs has now been linked to signaling through the TNF-like receptors RANK and CD40 (Akiyama et al. 2008; Rossi et al. 2007; Hikosaka et al. 2008) and to a lesser extent, signaling through the lymphotoxin beta receptor (Venanzi et al. 2007; Mouri et al. 2011; Boehm et al. 2003). Mice with deficiencies in RANK/RANK-ligand or CD40/CD40 ligand have been shown to have a defect in the generation of mTEC hi cells in the medulla. In addition, intracellular signaling molecules downstream of these receptors like

TRAF6 and Nik have also been implicated in the mTEC maturation process (Akiyama et al. 2005; Kajiura et al. 2004).

The molecular mechanisms by which Aire helps promote TSA expression are also an area of intense investigation. The PHD1 domain of Aire has been demonstrated by multiple groups to bind specifically to H3 histone tails that have the fourth lysine position unmethylated (H3K4Me0) (Org et al. 2008; Koh et al. 2010). This recognition links Aire to an epigenetic mechanism for identifying TSA genes to target. An extensive survey of proteins that are pulled down with Aire in transfected cells has demonstrated that AIRE interacts with proteins involved in DNA repair, particularly DNA-PK, and this may be another clue as to how TSA genes are targeted for induction of gene expression (Abramson et al. 2010). Finally, Aire has also been implicated in promoting transcriptional elongation of stalled RNA polymerase and this may be yet another component of how TSA expression is induced in mTECs (Oven et al. 2007; Giraud et al. 2012). Interestingly, it is important to note that Aire-deficient mTECs still display the ability of expressing some TSAs. This suggests that there may be other yet to be identified factors that allow mTECs to express TSAs.

Another important APC population resident to the thymic medulla are bone marrow-derived DCs. Thymic DCs can be broadly segregated into three distinct subsets: (1) plasmacytoid DCs (pDCs), which are CD11c Intermediate, $B220^+$, (2) conventional DCs (cDCs), which are $CD11c^+$, $CD8^-$, $Sirp\alpha^+$, and (3) $CD8^+$ DCs, which are $CD11c^+$, $CD8^+$, and $Sirp\alpha^-$. Interestingly, the $CD8^+$ DC subset is thought to be derived from a pre-T lymphoid progenitor recruited into the thymus, while the other two subsets likely migrate in from the periphery through alternative pathways. All three subsets are equipped with MHC class II and co-stimulatory ligand expression although this is somewhat less in the pDC subset. There has been evidence that the cDC subset can pick up peripheral antigens and bring them to the thymus for the induction of tolerance (Li et al. 2009; Bonasio et al. 2006). How abundant this process is for the induction of central tolerance remains to be determined, however, it is important to note that if extensive migration of peripheral self-antigens to thymus via DC pick up and migration was a dominant mechanism, it would make little sense for the generation of the TSA expression system to exist in the mTEC compartment. Thymic DCs likely play a critical role in the display of blood-borne and ubiquitously expressed self-antigens in the thymic medulla given that they are professionally equipped for this process. It is also important to note that although thymic DCs preferentially home to the medulla, there are also detectable DCs in the cortex and thus, DCs are likely contributing to negative selection in both compartments.

DCs exhibit the ability to endocytose or phagocytose nearby antigens for presentation in the MHC class II pathway and also in the MHC Class I pathway via cross-presentation. Thus, one interesting prevailing model regarding the display of TSAs in the medulla is that mTECs may "hand off" such antigens for display to thymocytes. In fact, there is experimental data to support this notion, particularly for MHC class II restricted display (Gallegos and Bevan 2004; Taniguchi et al. 2012; Koble and Kyewski 2009), however, there is some conflicting data on this model (Hubert et al. 2011). Interestingly, mTECs are equipped with MHC class II expression and antigen processing and presentation machinery. This begs the question as to if and how TSAs are displayed directly on mTEC MHC class II since the preferred pathway for presentation is through exogenous acquisition of antigen. Recently, Klein and colleagues have demonstrated that at least some mTECs display evidence of autophagosomes and this may be a method for mTECs to have mTEC-derived TSAs crossover into the MHC Class II antigen presentation pathway (Nedjic et al. 2008). Taken together, a picture is beginning to emerge for central tolerance whereby specialized activities of various APC populations resident to the thymus work in concert to drive the delicate balance of positive and negative selection. The thymic medulla appears to be a dynamic compartment characterized by mTEC turnover and diverse self-antigen expression and display which helps promote the deletion of autoreactive specificities.

1.2 Tolerance Through Treg Selection

Regulatory T cells (Treg) are a subpopulation of T cells necessary for suppression of self-reactive T cells in the periphery. In addition to negative selection of autoreactive T cells, selection of Tregs occurs in the thymus. Both of these processes are critical for immune tolerance. An initial piece of evidence to suggest the presence of a suppressive T cell population that was derived from the thymus came from the observation of autoimmune pathology in neonatal thymectomized mice. Early studies by Nishizuka and Sakakura first described "ovarian dysgenesis" after thymectomy day 2 after birth but not day 4 (Nishizuka and Sakakura 1969). Although these results were initially thought to be due to a hormone deficiency, later experiments found it to be an autoimmune lesion and thus, first implicated the thymus as a critical organ for prevention of self-reactivity. The thymic emigrants necessary to prevent the pathology were later named suppressor T cells as introduction of T cells from adult mice prevented the autoimmunity caused by early thymectomy (Sakaguchi et al. 1982). This work and others suggested the presence of a T cell population present in the periphery with self-reactive specificity that was responsible for the autoimmune pathology observed in the thymectomized mice and another thymic-derived T cell subset that could suppress these cells. It is now accepted that conventional T cells also have the ability to differentiate into FoxP3-expressing Tregs in the periphery but this section of the review will focus on the differentiation of natural or thymic Tregs.

Tregs were later defined as a population of $CD4^+$ as well as $CD25^+$ ($IL\text{-}2R\alpha$ chain) cells (Sakaguchi et al. 1995). In these studies, depletion of the $CD25^+$ Treg subset from $CD4^+$ cells and subsequent transfer into nude mice resulted in autoimmune pathology. Further insight into the molecular determinants of Treg differentiation came from identifying a frameshift mutation in FoxP3 as the cause of the lymphoproliferative disorder in the Scurfy mouse phenotype (Brunkow et al. 2001). In parallel, human studies in Immunodysregulation Polyendocrinopathy

Enteropathy X-linked syndrome (IPEX) patients linked the Scurfy mouse phenotype with the pathologies observed in those patients and ultimately the FoxP3 mutation (Chatila et al. 2000; Wildin et al. 2001; Bennett et al. 2001). Soon after, FoxP3 was identified as the master regulator of Treg differentiation as deletion of FoxP3 in T cells caused autoimmune pathology and transfer of CD4$^+$ CD25$^+$ cells into FoxP3-deficient mice rescues from disease (Fontenot et al. 2005; Fontenot et al. 2003; Hori 2003; Khattri et al. 2003).

Treg TCR and self-antigen specificity. Evidence that TCR specificity is important for Treg differentiation came from the observation that not all specificities can generate Tregs. For example, Tregs were only detected in D011.10 TCR transgenic mice when there were endogenous TCR alpha chains present and not in the RAG knockout background (Itoh et al. 1999). The hypothesis that Tregs have specificity for self-antigen was supported by elegant studies where preferential selection of Tregs was observed only when both a TCR transgene and its cognate neo-self-antigen were present (Jordan et al. 2001). These and other studies using TCR transgenics and a second transgene as a source of cognate endogenous self-antigen formed the basis for the idea that the TCR avidity for self-antigen needed for Treg selection is intermediate between what drives positive and negative selection (Fig. 2a) (Knoechel et al. 2005). In support of the notion that the antigen specificity for Tregs is distinct from that of conventional T cells, TCR repertoire analysis of both thymic T cell populations showed that they do indeed have different repertoires although some overlap was observed (Hsieh et al. 2004; Pacholczyk et al. 2006; Wong et al. 2007). Retroviral expression in RAG-deficient TCR transgenic T cells of TCRa chains derived from CD25$^+$ but not CD25$^-$ T cells led to a high frequency of expansion in lymphopenic hosts (Hsieh et al. 2004). Again, these data provide further evidence for the self-antigen specificity of thymic Tregs.

Discovery of the Niche. Transgenic expression of TCRs derived from naturally arising Tregs led to the unexpected result of the lack of thymic Tregs (Bautista et al. 2009; Leung et al. 2009). Further analyses demonstrated that the frequency of Tregs was inversely proportional to the clonal frequency and the most efficient Treg generation with observed with very low precursor numbers. In addition, these studies showed that the number of Tregs plateaued suggesting that there is an intraclonal competition for limiting factors in the thymus. Retrogenic mice expressing thymic Treg TCRs proliferate only under lymphopenic conditions presumably when antigen is no longer limiting (Hsieh et al. 2004, 2006). This 'niche' model suggests that whether or not a T cell becomes a Treg depends on what interactions it has in the thymus, which is distinct from clonal deletion.

TCR signal strength. Contrary to Brunet's clonal selection theory, encounter with self-antigen in the thymus has two outcomes: negative selection and Treg selection. An avidity-dependent selection process for Tregs was suggested by studies where lower affinity TCR transgenic thymocytes were not selected to become Tregs, even with expression of cognate antigen expressed on a second transgene (Jordan et al. 2001). Through the use of Nur77 GFP transgenic mice where GFP levels reflect TCR signal strength, higher GFP levels were detected in

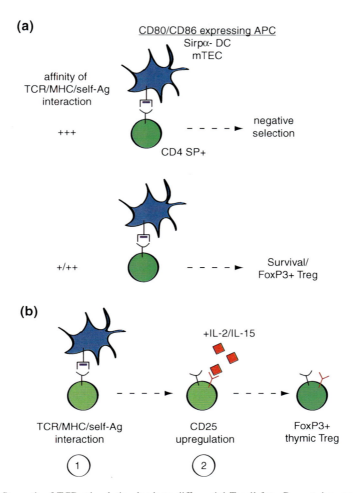

Fig. 2 a Strength of TCR stimulation leads to differential T cell fate. Recent data suggests that strong TCR stimulation induces negative selection, whereas moderate and lower affinity interactions lead to survival and/or regulatory T cell (Treg) differentiation. b Model for Treg development. Following a high/moderate affinity MHCII/self-antigen interaction with the Treg TCR, the Treg precursor then upregulates CD25. CD25 allows responsiveness to IL-2-mediated signals which induce expression of FoxP3 and lead to the generation of a mature regulatory T cell

thymic Tregs. In a Treg-derived TCR transgenic also containing the Nur77 reporter, higher GFP levels were only detected with lower precursor frequency (Moran et al. 2011). The importance of signal strength may explain why certain Treg TCR transgenics may result in negative selection as transgene expression levels may favor deletion (DiPaolo and Shevach 2009). Another scenario that suggests that moderate affinity is important for Treg induction is the reduction of antigen presentation in mTECs through synthetic miRNA-mediated knockdown of CIITA. In these studies, T cell fate was redirected from deletion to Treg cell selection (Hinterberger et al. 2010). Recent studies using the RIP-mOva and TCR retrogenics that recognize Ova with varying affinity suggest that the affinity needed for negative selection is about 100-fold higher than that found to promote Treg differentiation (Lee et al. 2012). In addition, those TCRs that led to Treg generation ranged in affinity within a 1000-fold range.

Cytokine plus self-antigen As mentioned earlier, Tregs express the marker CD25 on their cell surface, which also suggests continued stimulation from antigen encounter in the thymus. Aside from potentially indicating thymic Treg activation, expression of CD25 or the alpha chain of the high affinity IL-2 receptor on Tregs implicated IL-2 signaling in Treg maintenance. Evidence of the importance of IL-2 in Treg generation was shown through the observed autoimmunity in IL-2 deficient mice (Sadlack et al. 1993; Schorle et al. 1991). Similarly, the autoimmune phenotype of CD25 or CD122 (IL-2R beta chain) knockout mice reinforces the idea that IL-2 signaling is necessary for thymic Treg maintenance (Suzuki et al. 1995; Willerford et al. 1995). Although the phenotype of IL-2 deficient mice is less severe than what has been observed in STAT5-deficient animals, it is possible that other cytokines that share the common gamma chain such as IL-7 and IL-15 also contribute to Treg maintenance in the absence of IL-2 (Yao et al. 2007). From these studies and others, the following two-step model has been proposed where TCR engagement leads to the upregulation of IL-2R on thymic Tregs and consequently FoxP3 expression (Fig. 2b) (Lio and Hsieh 2008). A model in which STAT5 was constitutively expressed resulted in a larger Treg compartment (Burchill et al. 2008). IL-2 signaling may also influence Tregs through shaping the repertoire composition and size or by providing survival signals for Tregs (Burchill et al. 2008; D'Cruz and Klein 2005).

Additionally, TGFβ signaling has been implicated in thymic Treg development in that both TGFβRI and TGFβRII deficient mice had a reduction in the first wave of thymic Tregs (Liu et al. 2008; Marie et al. 2006). Although initial studies in TGFβRII knockout mice had no observed difference in thymic Tregs, it is thought that perhaps IL-2 induced proliferation of the remaining Tregs was responsible for the recovery in Treg numbers (Ouyang et al. 2010).

Cell types that induce Treg differentiation. The antigen presenting cells in the thymus also contribute to Treg selection. Various groups have shown the importance of CD28/B7.1 and B7.2 signaling for nTreg induction and all thymic APC types can contribute to this costimulatory signal. Some of the first evidence in support of mTECs being important in nTreg selection came from studies showing that a self-antigen expression by the thymic stroma induced the generation of Tregs (Jordan et al. 2001; Apostolou et al. 2002). Transgenic expression of hemagglutinin (HA) under the control of Aire regulatory elements along with an HA-specific TCR showed that expression of antigen specifically in the thymic epithelium resulted in Treg generation (Aschenbrenner et al. 2007). Furthermore, studies showing that lowering MHCII levels specifically in mTECs using transgenic expression of a microRNA specific to CIITA leads to enhanced Treg selection suggest that direct antigen presentation by mTECs leads to efficient Treg selection (Hinterberger et al. 2010). mTECs may also serve as a self-antigen reservoir for DCs that influence Treg differentiation. Both pDCs and cDCs can be found in the thymus. DCs have been shown to have a strong capacity to induce in vitro Treg differentiation with the $CD8^{lo}$ Sirpa$^+$ subset being most robust (Watanabe et al. 2005; Proietto et al. 2008). Conversely, depletion of DCs using the CD11cDTR transgenic mouse led to a reduction in Treg numbers (Darrasse-Jeze et al. 2009). It has been proposed that both DCs and mTECs cooperate to eliminate autoreactive T cells and similarly, this may be the case during Treg differentiation (Gallegos and Bevan 2004; Spence and Green 2008). Further, epithelial cells may provide cues for thymic DCs like XCL-1 for DC migration or TSLP for tDC expression of costimulatory molecules (Watanabe et al. 2005; Lei et al. 2011). Aside from being involved in the differentiation of Tregs, it remains to be tested if mTECs influence Tregs in other aspects such as TCR repertoire.

2 Concluding Remarks

As the site critical for the education of T cells, the thymus is necessary to limit the escape of self-reactive conventional CD4$^+$ T cells to the periphery. The thymus is also essential to generate regulatory T cells that suppress conventional CD4$^+$ T cells and thus impose additional tolerance mechanisms. Remarkably, the interactions of thymic APCs and T cells likely follows a complex series of events that include positive selection in early T cell development with subsequent negative selection or positive Treg selection of T cell clones with autoreactive specificities and this has been an area that has seen rapid progress in recent years. This elegant selection process thus allows for a diverse T cell repertoire that is tolerant to self, yet poised for immune responses against diverse pathogens. Moving forward, future challenges lie on how this selection process could be manipulated to either enhance tolerance as in autoimmune disease settings or break tolerance as in cancer immunotherapy.

Acknowledgments This work was supported by the NIH, The Helmsley Charitable Trust, and The Burroughs Wellcome Fund. The authors have no conflict of interest to disclose.

References

Aaltonen J (1997) An autoimmune disease, APECED, caused by mutations in a novel gene featuring two PHD-type zinc-finger domains. Nat Genet 17(4):399–403. doi:10.1038/ng1297-399

Abramson J, Giraud M, Benoist C, Mathis D (2010) Aire's partners in the molecular control of immunological tolerance. Cell 140(1):123–135. doi:10.1016/j.cell.2009.12.030, S0092-8674(09)01616-X [pii]

Akiyama T, Maeda S, Yamane S, Ogino K, Kasai M, Kajiura F, Matsumoto M, Inoue J-I (2005) Dependence of self-tolerance on TRAF6-directed development of thymic stroma. Science 308:248–251

Akiyama T, Shimo Y, Yanai H, Qin J, Ohshima D, Maruyama Y, Asaumi Y, Kitazawa J, Takayanagi H, Penninger JM, Matsumoto M, Nitta T, Takahama Y, Inoue J-I (2008) The tumor necrosis factor family receptors RANK and CD40 cooperatively establish the thymic medullary microenvironment and self-tolerance. Immunity 29:423–437

Anderson MS, Venanzi ES, Klein L, Chen Z, Berzins SP, Turley SJ, von Boehmer H, Bronson R, Dierich A, Benoist C, Mathis D (2002) Projection of an immunological self shadow within the thymus by the aire protein. Science 298:1395–1401

Apostolou I, Sarukhan A, Klein L, von Boehmer H (2002) Origin of regulatory T cells with known specificity for antigen. Nat Immunol 3:756–763

Aschenbrenner K, D'Cruz LM, Vollmann EH, Hinterberger M, Emmerich J, Swee LK, Rolink A, Klein L (2007) Selection of Foxp3+ regulatory T cells specific for self antigen expressed and presented by Aire+ medullary thymic epithelial cells. Nat Immunol 8:351–358

Baldwin TA, Sandau MM, Jameson SC, Hogquist KA (2005) The timing of TCR alpha expression critically influences T cell development and selection. J Exp Med 202(1):111–121. doi:10.1084/jem.20050359, jem.20050359 [pii]

Bautista JL, Lio C-WJ, Lathrop SK, Forbush K, Liang Y, Luo J, Rudensky AY, Hsieh C-S (2009) Intraclonal competition limits the fate determination of regulatory T cells in the thymus. Nat Immunol 10:610–617

Bennett CL, Christie J, Ramsdell F, Brunkow ME, Ferguson PJ, Whitesell L, Kelly TE, Saulsbury FT, Chance PF, Ochs HD (2001) The immune dysregulation, polyendocrinopathy, enteropathy, X-linked syndrome (IPEX) is caused by mutations of FOXP3. Nat Genet 27(1):20–21. doi:10.1038/83713

Boehm T, Scheu S, Pfeffer K, Bleul CC (2003) Thymic medullary epithelial cell differentiation, thymocyte emigration, and the control of autoimmunity require lympho-epithelial cross talk via LTbetaR. J Exp Med 198:757–769

Bonasio R, Scimone ML, Schaerli P, Grabie N, Lichtman AH, von Andrian UH (2006) Clonal deletion of thymocytes by circulating dendritic cells homing to the thymus. Nat Immunol 7:1092–1100

Bouillet P, Purton JF, Godfrey DI, Zhang LC, Coultas L, Puthalakath H, Pellegrini M, Cory S, Adams JM, Strasser A (2002) BH3-only Bcl-2 family member Bim is required for apoptosis of autoreactive thymocytes. Nature 415(6874):922–926. doi:10.1038/415922a, 415922a [pii]

Brunkow ME, Jeffery EW, Hjerrild KA, Paeper B, Clark LB, Yasayko SA, Wilkinson JE, Galas D, Ziegler SF, Ramsdell F (2001) Disruption of a new forkhead/winged-helix protein, scurfin, results in the fatal lymphoproliferative disorder of the scurfy mouse. Nat Genet 27:68–73

Burchill MA, Yang J, Vang KB, Moon JJ, Chu HH, Lio C-WJ, Vegoe AL, Hsieh C-S, Jenkins MK, Farrar MA (2008) Linked T cell receptor and cytokine signaling govern the development of the regulatory T cell repertoire. Immunity 28:112–121

Burnet F (1958) The clonal selection theory of acquired immunity. Vanderbilt University Press, Nashville

Calnan BJ, Szychowski S, Chan FK, Cado D, Winoto A (1995) A role for the orphan steroid receptor Nur77 in apoptosis accompanying antigen-induced negative selection. Immunity 3(3):273–282

Chatila TA, Blaeser F, Ho N, Lederman HM, Voulgaropoulos C, Helms C, Bowcock AM (2000) JM2, encoding a fork head-related protein, is mutated in X-linked autoimmunity-allergic disregulation syndrome. J Clin Invest 106:75–81

Cho HJ, Edmondson SG, Miller AD, Sellars M, Alexander ST, Somersan S, Punt JA (2003) Cutting edge: identification of the targets of clonal deletion in an unmanipulated thymus. J Immunol 170(1):10–13

D'Cruz LM, Klein L (2005) Development and function of agonist-induced CD25+ Foxp3+ regulatory T cells in the absence of interleukin 2 signaling. Nat Immunol 6:1152–1159

Daniels MA, Teixeiro E, Gill J, Hausmann B, Roubaty D, Holmberg K, Werlen G, Hollander GA, Gascoigne NR, Palmer E (2006) Thymic selection threshold defined by compartmentalization of Ras/MAPK signalling. Nature 444(7120):724–729. doi:10.1038/nature05269, nature05269 [pii]

Darrasse-Jeze G, Deroubaix S, Mouquet H, Victora GD, Eisenreich T, Yao Kh, Masilamani RF, Dustin ML, Rudensky A, Liu K, Nussenzweig MC (2009) Feedback control of regulatory T cell homeostasis by dendritic cells in vivo. J Exp Med 206:1853–1862

Derbinski J, Gabler J, Brors B, Tierling S, Jonnakuty S, Hergenhahn M, Peltonen L, Walter J, Kyewski B (2005) Promiscuous gene expression in thymic epithelial cells is regulated at multiple levels. J Exp Med 202(1):33–45. doi:10.1084/jem.20050471, jem.20050471 [pii]

Derbinski J, Pinto S, Rosch S, Hexel K, Kyewski B (2008) Promiscuous gene expression patterns in single medullary thymic epithelial cells argue for a stochastic mechanism. Proc Natl Acad Sci USA 105(2):657–662. doi:10.1073/pnas.0707486105, 0707486105 [pii]

Derbinski J, Schulte A, Kyewski B, Klein L (2001) Promiscuous gene expression in medullary thymic epithelial cells mirrors the peripheral self. Nat Immunol 2(11):1032–1039. doi:10.1038/ni723ni723, ni723 [pii]

DeVoss JJ, LeClair NP, Hou Y, Grewal NK, Johannes KP, Lu W, Yang T, Meagher C, Fong L, Strauss EC, Anderson MS (2010) An autoimmune response to odorant binding protein 1a is associated with dry eye in the Aire-deficient mouse. J Immunol 184:4236–4246

DiPaolo RJ, Shevach EM (2009) CD4+ T-cell development in a mouse expressing a transgenic TCR derived from a Treg. Eur J Immunol 39:234–240

Fassett MS, Jiang W, D'Alise AM, Mathis D, Benoist C (2012) Nuclear receptor Nr4a1 modulates both regulatory T-cell (Treg) differentiation and clonal deletion. Proc Natl Acad Sci USA 109(10):3891–3896. doi:10.1073/pnas.1200090109, 1200090109 [pii]

Fehling HJ, Krotkova A, Saint-Ruf C, von Boehmer H (1995) Crucial role of the pre-T-cell receptor alpha gene in development of alpha beta but not gamma delta T cells. Nature 375(6534):795–798. doi:10.1038/375795a0

Fontenot JD, Gavin MA, Rudensky AY (2003) Foxp3 programs the development and function of CD4+ CD25+ regulatory T cells. Nat Immunol 4(4):330–336. doi:10.1038/ni904, ni904 [pii]

Fontenot JD, Rasmussen JP, Williams LM, Dooley JL, Farr AG, Rudensky AY (2005) Regulatory T cell lineage specification by the forkhead transcription factor foxp3. Immunity 22:329–341

Fowlkes BJ, Schwartz RH, Pardoll DM (1988) Deletion of self-reactive thymocytes occurs at a CD4+8+ precursor stage. Nature 334(6183):620–623. doi:10.1038/334620a0

Gallegos AM, Bevan MJ (2004) Central tolerance to tissue-specific antigens mediated by direct and indirect antigen presentation. J Exp Med 200:1039–1049

Gil D, Schrum AG, Alarcon B, Palmer E (2005) T cell receptor engagement by peptide-MHC ligands induces a conformational change in the CD3 complex of thymocytes. J Exp Med 201(4):517–522. doi:10.1084/jem.20042036, jem.20042036 [pii]

Giraud M, Yoshida H, Abramson J, Rahl PB, Young RA, Mathis D, Benoist C (2012) Aire unleashes stalled RNA polymerase to induce ectopic gene expression in thymic epithelial cells. Proc Natl Acad Sci USA 109:535–540

Gray D, Abramson J, Benoist C, Mathis D (2007) Proliferative arrest and rapid turnover of thymic epithelial cells expressing Aire. J Exp Med 204:2521–2528

Gray DHD, Kupresanin F, Berzins SP, Herold MJ, O'Reilly LA, Bouillet P, Strasser A (2012) The BH3-only proteins Bim and Puma cooperate to impose deletional tolerance of organ-specific antigens. Immunity 37:451–462

Hikosaka Y, Nitta T, Ohigashi I, Yano K, Ishimaru N, Hayashi Y, Matsumoto M, Matsuo K, Penninger JM, Takayanagi H, Yokota Y, Yamada H, Yoshikai Y, Inoue J-I, Akiyama T, Takahama Y (2008) The cytokine RANKL produced by positively selected thymocytes fosters medullary thymic epithelial cells that express autoimmune regulator. Immunity 29:438–450

Hinterberger M, Aichinger M, da Costa OP, Voehringer D, Hoffmann R, Klein L (2010) Autonomous role of medullary thymic epithelial cells in central CD4(+) T cell tolerance. Nat Immunol 11:512–519

Hori S (2003) Control of regulatory T cell development by the transcription factor Foxp3. Science 299:1057–1061

Hsieh C-S, Liang Y, Tyznik AJ, Self SG, Liggitt D, Rudensky AY (2004) Recognition of the peripheral self by naturally arising CD25+ CD4+ T cell receptors. Immunity 21:267–277

Hsieh C-S, Zheng Y, Liang Y, Fontenot JD, Rudensky AY (2006) An intersection between the self-reactive regulatory and nonregulatory T cell receptor repertoires. Nat Immunol 7:401–410

Hubert FX, Kinkel SA, Davey GM, Phipson B, Mueller SN, Liston A, Proietto AI, Cannon PZF, Forehan S, Smyth GK, Wu L, Goodnow CC, Carbone FR, Scott HS, Heath WR (2011) Aire regulates the transfer of antigen from mTECs to dendritic cells for induction of thymic tolerance. Blood 118:2462–2472

Itoh M, Takahashi T, Sakaguchi N, Kuniyasu Y, Shimizu J, Otsuka F, Sakaguchi S (1999) Thymus and autoimmunity: production of CD25+ CD4+ naturally anergic and suppressive T cells as a key function of the thymus in maintaining immunologic self-tolerance. J Immunol 162:5317–5326

Jordan MS, Boesteanu A, Reed AJ, Petrone AL, Holenbeck AE, Lerman MA, Naji A, Caton AJ (2001) Thymic selection of CD4+ CD25+ regulatory T cells induced by an agonist self-peptide. Nat Immunol 2:301–306

Kajiura F, Sun S, Nomura T, Izumi K, Ueno T, Bando Y, Kuroda N, Han H, Li Y, Matsushima A, Takahama Y, Sakaguchi S, Mitani T, Matsumoto M (2004) NF-kappa B-inducing kinase establishes self-tolerance in a thymic stroma-dependent manner. J Immunol 172:2067–2075

Kappler JW, Roehm N, Marrack P (1987) T cell tolerance by clonal elimination in the thymus. Cell 49(2):273–280, 0092-8674(87)90568-X [pii]

Khattri R, Cox T, Yasayko S-A, Ramsdell F (2003) An essential role for Scurfin in CD4+ CD25+ T regulatory cells. Nat Immunol 4:337–342

Kisielow P, Bluthmann H, Staerz UD, Steinmetz M, von Boehmer H (1988) Tolerance in T-cell-receptor transgenic mice involves deletion of nonmature CD4+ 8+ thymocytes. Nature 333(6175):742–746. doi:10.1038/333742a0

Knoechel B, Lohr J, Kahn E, Bluestone JA, Abbas AK (2005) Sequential development of interleukin 2-dependent effector and regulatory T cells in response to endogenous systemic antigen. J Exp Med 202:1375–1386

Koble C, Kyewski B (2009) The thymic medulla: a unique microenvironment for intercellular self-antigen transfer. J Exp Med 206:1505–1513

Koh AS, Kingston RE, Benoist C, Mathis D (2010) Global relevance of Aire binding to hypomethylated lysine-4 of histone-3. Proc Natl Acad Sci USA 107(29):13016–13021. doi:10.1073/pnas.1004436107, 1004436107 [pii]

Kwan J, Killeen N (2004) CCR7 directs the migration of thymocytes into the thymic medulla. J Immunol 172(7):3999–4007

Lee H-M, Bautista JL, Scott-Browne J, Mohan JF, Hsieh C-S (2012) A broad range of self-reactivity drives thymic regulatory T cell selection to limit responses to self. Immunity 37:475–486

Lei Y, Ripen AM, Ishimaru N, Ohigashi I, Nagasawa T, Jeker LT, Bosl MR, Hollander GA, Hayashi Y, De Waal Malefyt R, Nitta T, Takahama Y (2011) Aire-dependent production of XCL1 mediates medullary accumulation of thymic dendritic cells and contributes to regulatory T cell development. J Exp Med 208:383–394

Leung MWL, Shen S, Lafaille JJ (2009) TCR-dependent differentiation of thymic Foxp3+ cells is limited to small clonal sizes. J Exp Med 206:2121–2130

Li J, Park J, Foss D, Goldschneider I (2009) Thymus-homing peripheral dendritic cells constitute two of the three major subsets of dendritic cells in the steady-state thymus. J Exp Med 206(3):607–622. doi:10.1084/jem.20082232, jem.20082232 [pii]

Lio C-WJ, Hsieh C-S (2008) A two-step process for thymic regulatory T cell development. Immunity 28:100–111

Liu Y, Zhang P, Li J, Kulkarni AB, Perruche S, Chen W (2008) A critical function for TGF-beta signaling in the development of natural CD4+ CD25+ Foxp3+ regulatory T cells. Nat Immunol 9:632–640

Marie JC, Liggitt D, Rudensky AY (2006) Cellular mechanisms of fatal early-onset autoimmunity in mice with the T cell-specific targeting of transforming growth factor-beta receptor. Immunity 25(3):441–454. doi:10.1016/j.immuni.2006.07.012, S1074-7613(06)00388-8 [pii]

McCaughtry TM, Wilken MS, Hogquist KA (2007) Thymic emigration revisited. J Exp Med 204(11):2513–2520. doi:10.1084/jem.20070601, S1074-7613(06)00388-8 [pii]

McNeil LK, Starr TK, Hogquist KA (2005) A requirement for sustained ERK signaling during thymocyte positive selection in vivo. Proc Natl Acad Sci USA 102(38):13574–13579. doi:10.1073/pnas.0505110102, 0505110102 [pii]

Metzger TC, Anderson MS (2011) Control of central and peripheral tolerance by Aire. Immunol Rev 241:89–103

Moran AE, Holzapfel KL, Xing Y, Cunningham NR, Maltzman JS, Punt J, Hogquist KA (2011) T cell receptor signal strength in Treg and iNKT cell development demonstrated by a novel fluorescent reporter mouse. J Exp Med 208:1279–1289

Mouri Y, Yano M, Shinzawa M, Shimo Y, Hirota F, Nishikawa Y, Nii T, Kiyonari H, Abe T, Uehara H, Izumi K, Tamada K, Chen L, Penninger JM, Inoue J-I, Akiyama T, Matsumoto M (2011) Lymphotoxin signal promotes thymic organogenesis by eliciting RANK expression in the embryonic thymic stroma. J Immunol 186:5047–5057

Murata S, Sasaki K, Kishimoto T, Niwa S, Hayashi H, Takahama Y, Tanaka K (2007) Regulation of CD8+ T cell development by thymus-specific proteasomes. Science 316(5829):1349–1353. doi:10.1126/science.1141915, 316/5829/1349 [pii]

Murphy KM, Heimberger AB, Loh DY (1990) Induction by antigen of intrathymic apoptosis of CD4+ CD8+ TCRlo thymocytes in vivo. Science 250(4988):1720–1723

Nedjic J, Aichinger M, Emmerich J, Mizushima N, Klein L (2008) Autophagy in thymic epithelium shapes the T-cell repertoire and is essential for tolerance. Nature 455:396–400

Nika K, Soldani C, Salek M, Paster W, Gray A, Etzensperger R, Fugger L, Polzella P, Cerundolo V, Dushek O, Hofer T, Viola A, Acuto O (2010) Constitutively active Lck kinase in T cells drives antigen receptor signal transduction. Immunity 32(6):766–777. doi:10.1016/j.immuni.2010.05.011, S1074-7613(10)00203-7 [pii]

Nishizuka Y, Sakakura T (1969) Thymus and reproduction: sex-linked dysgenesis of the gonad after neonatal thymectomy in mice. Science 166(3906):753–755

Oettinger MA, Schatz DG, Gorka C, Baltimore D (1990) RAG-1 and RAG-2, adjacent genes that synergistically activate V(D)J recombination. Science 248(4962):1517–1523

Org T, Chignola F, Hetenyi C, Gaetani M, Rebane A, Liiv I, Maran U, Mollica L, Bottomley MJ, Musco G, Peterson P (2008) The autoimmune regulator PHD finger binds to non-methylated histone H3K4 to activate gene expression. EMBO Rep 9(4):370–376. doi:10.1038/sj.embor.2008.11 embor200811 [pii]

Ouyang W, Beckett O, Ma Q, Li MO (2010) Transforming growth factor-beta signaling curbs thymic negative selection promoting regulatory T cell development. Immunity 32: 642–653

Oven I, Brdickova N, Kohoutek J, Vaupotic T, Narat M, Peterlin BM (2007) AIRE recruits P-TEFb for transcriptional elongation of target genes in medullary thymic epithelial cells. Mol Cell Biol 27(24):8815–8823. doi:10.1128/MCB.01085-07, MCB.01085-07 [pii]

Pacholczyk R, Ignatowicz H, Kraj P, Ignatowicz L (2006) Origin and T cell receptor diversity of Foxp3+ CD4+ CD25+ T cells. Immunity 25:249–259

Proietto AI, van Dommelen S, Zhou P, Rizzitelli A, D'Amico A, Steptoe RJ, Naik SH, Lahoud MH, Liu Y, Zheng P, Shortman K, Wu L (2008) Dendritic cells in the thymus contribute to T-regulatory cell induction. Proc Natl Acad Sci USA 105:19869–19874

Rossi SW, Kim M-Y, Leibbrandt A, Parnell SM, Jenkinson WE, Glanville SH, McConnell FM, Scott HS, Penninger JM, Jenkinson EJ, Lane PJL, Anderson G (2007) RANK signals from CD4(+)3(−) inducer cells regulate development of Aire-expressing epithelial cells in the thymic medulla. J Exp Med 204:1267–1272

Sadlack B, Merz H, Schorle H, Schorle H, Schimpl A, Feller AC, Horak I (1993) Ulcerative colitis-like disease in mice with a disrupted interleukin-2 gene. Cell 75(2):253–261, 0092-8674(93)80067-O [pii]

Sakaguchi S, Sakaguchi N, Asano M, Itoh M, Toda M (1995) Immunologic self-tolerance maintained by activated T cells expressing IL-2 receptor alpha-chains (CD25). Breakdown of a single mechanism of self-tolerance causes various autoimmune diseases. J Immunol 155:1151–1164

Sakaguchi S, Takahashi T, Nishizuka Y (1982) Study on cellular events in post-thymectomy autoimmune oophoritis in mice. II. Requirement of Lyt-1 cells in normal female mice for the prevention of oophoritis. J Exp Med 156:1577–1586

Schatz DG, Oettinger MA, Baltimore D (1989) The V(D)J recombination activating gene, RAG-1. Cell 59(6):1035–1048. 0092-8674(89)90760-5 [pii]

Schorle H, Holtschke T, Hunig T, Schimpl A, Horak I (1991) Development and function of T cells in mice rendered interleukin-2 deficient by gene targeting. Nature 352(6336):621–624. doi:10.1038/352621a0

Shum AK, DeVoss J, Tan CL, Hou Y, Johannes K, O'Gorman CS, Jones KD, Sochett EB, Fong L, Anderson MS (2009) Identification of an autoantigen demonstrates a link between interstitial lung disease and a defect in central tolerance. Sci Transl Med 1(9):9ra20. doi:10.1126/scitranslmed.3000284, 1/9/9ra20 [pii]

Smith KM, Olson DC, Hirose R, Hanahan D (1997) Pancreatic gene expression in rare cells of thymic medulla: evidence for functional contribution to T cell tolerance. Int Immunol 9(9):1355–1365

Sohn SJ, Thompson J, Winoto A (2007) Apoptosis during negative selection of autoreactive thymocytes. Curr Opin Immunol 19(5):510–515. doi:10.1016/j.coi.2007.06.001, S0952-7915(07)00104-5 [pii]

Spence PJ, Green EA (2008) Foxp3+ regulatory T cells promiscuously accept thymic signals critical for their development. Proc Natl Acad Sci USA 105:973–978

Starr TK, Jameson SC, Hogquist KA (2003) Positive and negative selection of T cells. Annu Rev Immunol 21:139–176. doi:10.1146/annurev.immunol.21.120601.141107, 120601.141107 [pii]

Su MA, Giang K, Zumer K, Jiang H, Oven I, Rinn JL, DeVoss JJ, Johannes KPA, Lu W, Gardner J, Chang A, Bubulya P, Chang HY, Peterlin BM, Anderson MS (2008) Mechanisms of an autoimmunity syndrome in mice caused by a dominant mutation in Aire. J Clin Invest 118:1712–1726

Suzuki H, Kundig TM, Furlonger C, Wakeham A, Timms E, Matsuyama T, Schmits R, Simard JJ, Ohashi PS, Griesser H (1995) Deregulated T cell activation and autoimmunity in mice lacking interleukin-2 receptor beta. Science 268(5216):1472–1476

Taniguchi RT, DeVoss JJ, Moon JJ, Sidney J, Sette A, Jenkins MK, Anderson MS (2012) Detection of an autoreactive T-cell population within the polyclonal repertoire that undergoes distinct autoimmune regulator (Aire)-mediated selection. Proc Natl Acad Sci USA 109:7847–7852

Thompson J, Winoto A (2008) During negative selection, Nur77 family proteins translocate to mitochondria where they associate with Bcl-2 and expose its proapoptotic BH3 domain. J Exp Med 205(5):1029–1036. doi:10.1084/jem.20080101, jem.20080101 [pii]

Ueno T, Saito F, Gray DH, Kuse S, Hieshima K, Nakano H, Kakiuchi T, Lipp M, Boyd RL, Takahama Y (2004) CCR7 signals are essential for cortex-medulla migration of developing thymocytes. J Exp Med 200(4):493–505. doi:10.1084/jem.20040643 jem.20040643 [pii]

Venanzi ES, Gray DH, Benoist C, Mathis D (2007) Lymphotoxin pathway and Aire influences on thymic medullary epithelial cells are unconnected. J Immunol 179(9):5693–5700, 179/9/5693 [pii]

Villasenor J, Besse W, Benoist C, Mathis D (2008) Ectopic expression of peripheral-tissue antigens in the thymic epithelium: probabilistic, monoallelic, misinitiated. Proc Natl Acad Sci USA 105(41):15854–15859. doi:10.1073/pnas.0808069105, 0808069105 [pii]

Watanabe N, Wang Y-H, Lee HK, Ito T, Wang Y-H, Cao W, Liu Y-J (2005) Hassall's corpuscles instruct dendritic cells to induce CD4+ CD25+ regulatory T cells in human thymus. Nature 436:1181–1185

Wildin RS, Ramsdell F, Peake J, Faravelli F, Casanova JL, Buist N, Levy-Lahad E, Mazzella M, Goulet O, Perroni L, Bricarelli FD, Byrne G, McEuen M, Proll S, Appleby M, Brunkow ME (2001) X-linked neonatal diabetes mellitus, enteropathy and endocrinopathy syndrome is the human equivalent of mouse scurfy. Nat Genet 27:18–20

Willerford DM, Chen J, Ferry JA, Davidson L, Ma A, Alt FW (1995) Interleukin-2 receptor alpha chain regulates the size and content of the peripheral lymphoid compartment. Immunity 3(4):521–530, 1074-7613(95)90180-9 [pii]

Wong J, Obst R, Correia-Neves M, Losyev G, Mathis D, Benoist C (2007) Adaptation of TCR repertoires to self-peptides in regulatory and nonregulatory CD4+ T cells. J Immunol 178:7032–7041

Yao Z, Kanno Y, Kerenyi M, Stephens G, Durant L, Watford WT, Laurence A, Robinson GW, Shevach EM, Moriggl R, Hennighausen L, Wu C, O'Shea JJ (2007) Nonredundant roles for Stat5a/b in directly regulating Foxp3. Blood 109:4368–4375

Yin X, Ladi E, Chan SW, Li O, Killeen N, Kappes DJ, Robey EA (2007) CCR7 expression in developing thymocytes is linked to the CD4 versus CD8 lineage decision. J Immunol 179(11):7358–7364, 179/11/7358 [pii]

Trafficking to the Thymus

Shirley L. Zhang and Avinash Bhandoola

Abstract The continuous production of T lymphocytes requires that hematopoietic progenitors developing in the bone marrow migrate to the thymus. Rare progenitors egress from the bone marrow into the circulation, then traffic via the blood to the thymus. It is now evident that thymic settling is tightly regulated by selectin ligands, chemokine receptors, and integrins, among other factors. Identification of these signals has enabled progress in identifying specific populations of hematopoietic progenitors that can settle the thymus. Understanding the nature of progenitor cells and the molecular mechanisms involved in thymic settling may allow for therapeutic manipulation of this process, and improve regeneration of the T lineage in patients with impaired T cell numbers.

Contents

1	Introduction	88
2	Candidate Thymic Settling Progenitors	89
3	Mobilization into the Circulation	90
4	Entering the Thymus	92
	4.1 The Molecular Basis of Homing	92
	4.2 The Thymic Settling Progenitor Niche	95
5	Fetal Versus Adult Thymic Settling	96
6	Evolution of the Thymus and Homing	98

S. L. Zhang · A. Bhandoola (✉)
Department of Pathology and Laboratory Medicine, Perelman School of Medicine,
University of Pennsylvania, 264 John Morgan Building 3620 Hamilton Walk,
Philadelphia, PA, USA
e-mail: bhandooa@mail.med.upenn.edu

Current Topics in Microbiology and Immunology (2014) 373: 87–111
DOI: 10.1007/82_2013_324
© Springer-Verlag Berlin Heidelberg 2013
Published Online: 27 April 2013

7 Regeneration of the T Lineage	99
8 Conclusions	101
References	102

1 Introduction

T lymphocytes are important in the adaptive immune response against pathogens, while preventing inappropriate immune response to self (Stritesky et al. 2012). The primary site of all T cell development is the thymus, which develops in the fetus from mesenchymal cells and epithelial cells that attract immature lymphocyte precursors (Anderson et al. 1993). T cell development begins in the fetus and continues throughout adulthood. In adults, T cells originate from bone marrow progenitors that egress from the bone marrow, traffic through the blood, and are imported into the thymus (Donskoy and Goldschneider 1992). Entry of progenitors into the thymus is regulated by the interaction of molecules on the surface of lymphoid precursors with ligands and receptors on the endothelial layer of blood vessels of the thymus. Progenitor cells immigrating to the thymus receive signals to divide, differentiate, and commit to the T cell fate. The T cell receptor (TCR) loci are recombined, and developing cells undergo positive and negative selection based on the specificity of their TCRs (Morris and Allen 2012). Several lineages of T cells mature in the thymus, including $\gamma\delta$ T cells, and subsets of $\alpha\beta$ T cells including NK-T cells, CD4 helper cells, and CD8 cytotoxic T cells. Finally, the matured T cells are released from the thymus into the periphery (Weinreich and Hogquist 2008). Production of naïve T cells is maintained by the thymus through life (Storek et al. 2003).

This chapter summarizes the current understanding of progenitor trafficking to the thymus, which in adult mice involves progenitors exiting the bone marrow, traveling though the circulation, homing to the thymus, and surviving in the thymic settling progenitor niche. We first discuss the progenitors that are the likely thymic settling progenitors that ultimately give rise to T cells. We address the mechanisms known to be involved in bone marrow egress to the circulation and homing to the thymus in the adult and the fetus. Most of what is known about thymic settling has been established in mice; however, molecules implicated in thymic homing in mice have orthologues that are required for T cell development in other model species including zebrafish and medaka, suggesting that thymic homing is similar not only among mammalian species, but also in other vertebrates (Boehm et al. 2003).

Finally, we review potential clinical applications of modulating the process of thymic settling in order to improve the regeneration of T cells. Following bone marrow transplant (BMT), the T cell compartment is slow to reconstitute, leaving patients immunocompromised (van Den Brink et al. 2013). Several aspects of T cell development are impacted by BMT; we discuss whether aberrant thymic settling may contribute to slower reconstitution.

2 Candidate Thymic Settling Progenitors

T cell potential is broadly distributed among primitive hematopoietic progenitor populations; many distinct progenitors can become T cells. Among these, a more restricted group of thymic settling progenitors also possesses the ability to mobilize from the blood and enter the thymus. The experimental systems used to study progenitor populations that settle the thymus involve intravenous transfers of large quantities of progenitors that can be recovered at early time points, but may not reflect physiological levels of progenitors in the blood. An additional complication is that the transferred cells may not settle the thymus immediately, but may first differentiate into distinct cells that then settle the thymus (Porritt et al. 2004; Schwarz and Bhandoola 2004; Perry et al. 2006; Serwold et al. 2009). All of the progenitors described below have T cell potential, as they will differentiate into T cells if placed in the thymus by intrathymic injection. The characteristics of thymic settling progenitors have been discussed in detail in several reviews (Boehm and Bleul 2006; Bhandoola et al. 2007; Iwasaki and Akashi 2007; Izon 2008; Koch and Radtke 2011; Zlotoff and Bhandoola 2011).

During fetal development, hematopoiesis occurs primarily in the liver, whereas in adults, candidate thymic settling progenitors reside within the bone marrow (Mikkola and Orkin 2006). Hematopoietic progenitors are functionally and phenotypically heterogeneous, and can be distinguished by self-renewal capacity, lineage potential, and combinations of surface markers. Hematopoietic stem cells (HSC) are the precursor cells from which all blood cell types, including T cells, originate. HSC have the ability to self-renew and continuously give rise to all blood cell types throughout the lifetime of the organism (Spangrude et al. 1988; Osawa et al. 1996). HSC develop via non-renewing progenitor intermediates into terminally differentiated blood cells, which have lost the potential to differentiate into any other lineages. Phenotypically, HSC are negative for all lineage markers and have high levels of Kit and Sca1 expression (LSK, for Lineage-negative, $Sca1^+$ Kit^+) (Table 1).

HSC give rise to multipotent progenitors (MPP), which can also reconstitute all blood lineages, but cannot self-renew (Adolfsson et al. 2001). MPP share the LSK phenotype with HSCs, but can be separated from HSC by surface expression of the cytokine receptor Flt3, among other markers (Morrison et al. 1997; Christensen and Weissman 2001; Wilson et al. 2008). MPP subsequently give rise to a subset of cells termed lymphoid-primed multipotent progenitors (LMPP), which are the highest Flt3-expressing LSK population. LMPP are also termed earliest lymphocyte progenitors (ELP) due to the expression of early lymphoid genes such as Recombination Activating Gene 1 (RAG1) and RAG2 (Igarashi et al. 2002). These cells retain myeloid and lymphoid potential, but no longer possess megakaryocyte or erythroid potential (Adolfsson et al. 2005).

Table 1 Candidate thymic settling progenitors

Candidate thymic settling progenitors	Surface phenotype	Lineage potential	Homing molecules
HSC	$Lin^- Sca1^+cKit^+Flt3^-$	All blood cells	PSGL-1
MPP	$Lin^-Sca1^+cKit^+Flt3^{lo}$	All blood cells	PSGL-1
LMPP	$Lin^-Sca1^+cKit^+Flt3^{hi}$	Lymphoid Myeloid	CCR7, CCR9 PSGL-1 (functional)
Ly6D$^-$CLP	$Lin^-Sca1^{lo}cKit^{lo}Flt3^{hi}$ $IL\text{-}7R^+Ly6D^-$	Lymphoid Myeloid (residual)	CCR7 PSGL-1 (functional)
Ly6D$^+$CLP	$Lin^-Sca1^{lo}cKit^{lo}Flt3^{hi}$ $IL\text{-}7R^+Ly6D^+B220^{-/+}$	Lymphoid	CCR7, CCR9, PSGL-1
ETP	$Lin^-Sca1^+cKit^+CD25^-$	Lymphoid Myeloid	N/A

Common lymphoid progenitors (CLP) differentiate from LMPP, accompanied by down-regulation of expression of Kit and Sca1, retention of high Flt3 expression, and the expression of high levels of surface IL-7 receptor (Kondo et al. 1997). CLP have lymphoid potential but greatly attenuated myeloid potential (Kondo et al. 1997). CLP are heterogenous, and can be further subdivided into the more primitive Ly6D$^-$ CLP and downstream Ly6D$^+$ CLP that also highly express RAG genes (Inlay et al. 2009; Mansson et al. 2011). Cells further downstream of Ly6D$^+$ CLP primarily produce B cells if they are retained in the bone marrow.

The rapid kinetics with which intravenously transferred LMPP and CLP develop into T cells suggest that physiological thymic settling progenitors are among these populations (Schwarz et al. 2007). Upstream populations such as HSC produce T lymphocytes with greatly delayed kinetics in comparison to LMPP and CLP, suggesting these upstream progenitors may home to the bone marrow prior to generating cells able to settle the thymus. In spite of these advances, the precise identity of the cells that traffic to the thymus has remained controversial (Allman et al. 2003; Martin et al. 2003; Porritt et al. 2004; Serwold et al. 2009). A method of narrowing down the pool of potential thymic settling progenitors is to determine the molecular requirements for trafficking from the bone marrow to the circulation and migration from the circulation to the thymus. The cellular machinery required for trafficking is only expressed on specific progenitor populations. As discussed later, this approach indicates that physiological thymus settling progenitors likely reside within LMPP and CLP subsets.

3 Mobilization into the Circulation

In order for adult bone marrow-derived progenitors to arrive at the thymus, progenitors with T lineage potential must first mobilize out of the bone marrow and into the blood. Cells with progenitor phenotypes have been detected in peripheral blood at low levels. Transfers of blood provide long-term multilineage reconstitution in recipients, indicating that functional HSCs circulate (Goodman and Hodgson 1962; Wright et al. 2001; Schwarz and Bhandoola 2004). MPP, LMPP, and CLP have been detected in the blood; however the very low number of progenitors in the circulation of adult mice has made it difficult to study the physiological regulation of bone marrow progenitor migration (Schwarz and Bhandoola 2004; Perry et al. 2006; Lai and Kondo 2007). For this reason, the signals that may govern physiological mobilization, if any, are largely unknown. However, there are pharmacological methods of mobilizing progenitors to the blood, which have suggested possible mechanisms of retention and egress. Exogenous administration of several cytokines and hormones, described below, can induce progenitor release from the bone marrow.

Bone marrow progenitors reside in localized niches along perivascular sites at the endosteum of the bone (Kunisaki and Frenette 2012). Many cells of the bone marrow are thought to support the HSC niche including osteoblasts, perivascular mesenchymal stem cells, and sinusoidal endothelial cells (Lymperi et al. 2010). Mesenchymal stem cells express adhesion molecules and chemokines, which retain and maintain hematopoietic progenitors within the bone marrow (Mendez-Ferrer et al. 2010). Vascular cell adhesion molecule 1 (VCAM-1) tethers stem and progenitor cells to the bone (Simmons et al. 1992). The chemokine receptor CXCR4 and its ligand CXCL12, also known as stromal cell derived factor-1, have been suggested to regulate the retention of progenitors (Pitchford et al. 2009). Blockade of CXCR4 through administration of an antagonist, AMD3100, allows for rapid mobilization of bone marrow progenitors into the blood. Other chemokines have been reported to mobilize progenitor cells into the circulation, including CCL3 (MIP-1α), CXCL8 (IL-8), and CXCL2 (GRO-β) (Marshall et al. 1997; King et al. 2001; Pruijt et al. 2002).

Erythropoietin, thrombopoietin, and granulocyte colony-stimulating factor (G-CSF) have each been found to induce bone marrow progenitor mobilization (Pettengell et al. 1994; Murray et al. 1998). G-CSF is widely used clinically to mobilize stem and progenitor cells for transplants. The response of HSC mobilization to G-CSF is more sustained than the response to AMD3100, although the effects of both therapies are transient. G-CSF is thought to act on the bone marrow niche through macrophage and neutrophil-mediated release of cytokines and metalloproteases that degrade the extracellular matrix (Liu et al. 2000; Pelus et al. 2004; Winkler et al. 2012). G-CSF has also been found to disrupt CXCL12 expression (Petit et al. 2002; Levesque et al. 2003). G-CSF and AMD3100 act synergistically to mobilize bone marrow progenitors (Broxmeyer et al. 2005).

More recently, sphingosine-1 phosphate (S1P) gradients have been found to be important in mediating progenitor egress from the bone marrow (Golan et al. 2012). When S1P receptors are inhibited with FTY720, hematopoietic progenitors are prevented from leaving the bone marrow and entering into the circulation. Furthermore, blockade of CXCR4 or administration of G-CSF has been found to upregulate S1P in blood plasma, which then facilitates progenitor egress (Ratajczak et al. 2010; Golan et al. 2012).

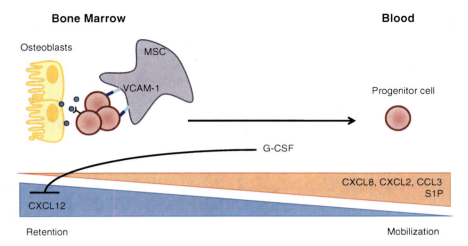

Fig. 1 Model for bone marrow progenitor mobilization. Progenitors in the bone marrow are retained by adhesion molecules such as VCAM-1 and the chemokine ligand CXCL12. Mobilization can be mediated through the increase in cytokines and S1P in blood or through G-CSF-induced downregulation of CXCL12

Together, these data suggest a model in which competing chemokine gradients favor either retention or mobility and thus regulate the rate of progenitor egress from the bone marrow (Fig. 1). This model would imply that at homeostasis, most progenitors in bone marrow receive more retention signals than mobilizing signals, but specific populations of progenitors may egress the bone marrow through the regulated control of their retention or mobilization molecules. Another possibility is that small numbers of progenitors randomly escape from the bone marrow into the blood. This latter hypothesis would be supported by equivalent relative ratios of different progenitor populations in the blood and the bone marrow.

4 Entering the Thymus

4.1 The Molecular Basis of Homing

Settling of the thymus by hematopoietic progenitors in adult mice is thought to be a rare event, and fewer than 20 cells per day have been estimated to lodge within the thymus (Wallis et al. 1975). This estimate is based on intravenous injection of a mixture of two bone marrow populations bearing congenic markers into irradiated mice. When the thymi of these mice were examined later, there was considerable variation in the ratio of the congenic markers expressed on thymocytes in different mice. Although this experiment involved radiation and is not the physiological condition, it indicates that the number of thymic settling progenitors can be very low. If this is also true in physiological conditions, this can help explain why it has been so difficult to identify progenitors within the thymus that have properties and surface phenotype identical to bone marrow populations.

The earliest defined T cell precursor population in the thymus are termed early thymic progenitors (ETP), which have been characterized as negative/low for CD4 and CD8 co-receptors (double negative cells; DN), positive for CD44 and Kit, but have not yet upregulated CD25 (DN1 Kit$^+$) (Shortman and Wu 1996; Allman et al. 2003; Porritt et al. 2004; Benz et al. 2008). The phenotype of the thymic settling progenitors has been shown to rapidly change once they enter the thymus and are exposed to Notch ligands (Krueger et al. 2006). Since putative rare upstream progenitors cannot yet be identified or analyzed directly, the characterization of ETP has been used to deduce the properties of thymic settling progenitors. As progenitor cells differentiate, developmental potentials for non-T cell lineages are lost; therefore, the more primitive thymic settling progenitors are those that have the same or more lineage potentials than ETP. ETP are not committed to the T cell lineage and subsets of ETP retain B cell, NK cell, dendritic cell, and a degree of myeloid potential (Bell and Bhandoola 2008; Luc et al. 2012). Some ETP adopt a granulocyte fate within the thymus, providing support for the idea that some thymic settling progenitors are myelo-lymphoid progenitors (Luc et al. 2012; De Obaldia et al. 2013).

Homing of progenitors to the thymus has been suggested to be similar to lymphocyte homing to lymph nodes, a process which involves selectin ligands, chemokine receptors and integrins (Scimone et al. 2006). In this model, progenitors are thought to undergo a multistep cascade, which includes: rolling adhesion; chemoattraction; tight adhesion; and transendothelial migration. L-selectin on T lymphocytes mediates rolling adhesion by binding to peripheral node addressin (PNAD) on the surface of endothelial cells. Expression of surface CCR7 allows lymphocytes to migrate toward the chemokine ligand CCL21, which is found on high endothelial venules. Signaling though the chemokine receptor activates integrins to acquire a high affinity structure. Integrin binding arrests the cell and additional signals induce transmigration of the cell through the endothelial layer into the lymph nodes (Springer 1994). In trafficking to the thymus, P-selectin glycoprotein ligand 1 (PSGL-1) and CCR7 or CCR9 on hematopoietic progenitors interact with targets on the endothelium for migration into the tissue (Rossi et al. 2005b; Scimone et al. 2006; Krueger et al. 2010; Zlotoff et al. 2010).

PSGL-1 is an important molecule present on bone marrow progenitors involved in homing to the thymus (Rossi et al. 2005b; Sultana et al. 2012). The receptor for PSGL-1 is P-selectin, which is present on the surface of some endothelial cells lining the vasculature of the thymus (Rossi et al. 2005b). Periodic alterations in levels of P-selectin on thymic endothelial cells are suggested to regulate the receptivity of the thymus to progenitors (Gossens et al. 2009). Progenitors deficient in PSGL-1 are inefficient at generating thymocytes in competition with wild-type progenitors; however, there is no deficiency in bone marrow progenitor generation (Rossi et al. 2005b). Although all bone marrow progenitors express

PSGL-1 transcripts, expression of functional protein able to bind P-selectin requires post-translational modifications via sialyation, glycosylation, and sulfation (Li et al. 1996). Expression of functional PSGL-1 is low in HSCs and MPPs, but upregulated on downstream CCR9$^+$ LMPP and Ly6D$^-$ CLP. Additionally, transcripts for glycosyltransferase enzymes Fut4 and Fut7, which are required for functional PSGL-1 expression, are most abundant in LMPP and Ly6D$^-$ CLP, and progenitors lacking Fut4 and Fut7 did not efficiently home to the thymus (Sultana et al. 2012).

The involvement of chemokine receptors in thymic homing was suggested by pertussis toxin blockade of G protein couple receptor (GPCR). Pertussis toxin treatment of bone marrow progenitors prior to intravenous transfer diminishes ETP numbers in the thymus, indicating that $G\alpha_i$ signaling in bone marrow progenitors is required for thymic homing (Jin and Wu 2008). The guanine nucleotide exchange factors (GEFs) involved in homing to the fetal thymus are DOCK2 and DOCK180. DOCK2 and DOCK180 double deficient mice have a smaller fetal thymus and progenitor cells lacking DOCK2 and DOCK180 are defective in in vitro migration to fetal thymic lobes; however, mutant progenitors were capable of producing T cells when cocultured on a fetal thymic stroma (Lei et al. 2009). Together, these results suggest that one or more receptors for chemoattractants (which are generally GPCRs) might be important in thymic homing.

Indeed, the chemokine receptors CCR7 and CCR9 have been implicated in thymic homing in fetal and adult mice (Uehara et al. 2002; Liu et al. 2006; Krueger et al. 2010; Zlotoff et al. 2010). The ligands for CCR7 and CCR9, which are CCL21/CCL19 and CCL25 respectively, are produced by stromal epithelial cells in the thymus (Ciofani and Zuniga-Pflucker 2007). CCL25 is expressed by all thymic epithelial cells (TEC) whereas CCL19 and CCL21 are expressed by medullary epithelial cells (Wurbel et al. 2000; Misslitz et al. 2004; Ueno et al. 2004). CCL25 is also found in the small intestine, whereas CCL21 is expressed in a wide range of organs (Vicari et al. 1997; Willimann et al. 1998). When progenitors deficient for either CCR7 and/or CCR9 are placed in competition with wild-type cells using mixed bone marrow irradiation chimeras, the knockout cells contribute less efficiently to thymic and peripheral T cell lineages; however, hematopoietic progenitor development in the bone marrow is largely unaffected. The defect is overcome when the cells are transferred directly to the thymus by intrathymic injection (Schwarz et al. 2007; Krueger et al. 2010; Zlotoff et al. 2010). These data together indicate that most efficient progenitor homing relies on CCR7 and CCR9. Other chemokines such as CXCR4 and CCR5 might also have a role in progenitor migration; however, the relative contribution of these receptors to homing in adult mice needs to be further evaluated (Robertson et al. 2006).

Integrins are thought to be responsible for mediating firm adhesion of rolling progenitors in the vasculature of the thymus (Ruiz et al. 1995). CLP express high levels of $\alpha L\beta 2$ and $\alpha 4\beta 1$ integrins, which bind the receptors ICAM-1 and VCAM-1 that are expressed by thymic endothelial cells. Inhibition of integrins on bone marrow progenitors transferred intravenously reduced trafficking to the thymus, suggesting that these integrin pairs play a role in homing (Scimone et al. 2006).

Although thymic settling progenitors have not been unambiguously identified, the available evidence implicates cells that express CCR9 or CCR7 (or both), functional PSGL-1, and integrins. HSC and MPP do not express either CCR9 or CCR7, eliminating them as efficient thymic homing progenitors. A subset of common myeloid progenitors possesses a degree of T cell potential in addition to myeloid potential; however, absence of CCR7 or CCR9 expression implies that normally these cells do not contribute to T lymphopoiesis (Chi et al. 2010). CCR9 and CCR7 are expressed on subsets of LMPP, and nearly all CLP express CCR7 and some also express CCR9. Furthermore, functional PSGL-1 is most highly expressed on $CCR9^+$ LMPP and $Ly6D^-$ CLP (Sultana et al. 2012). The gene expression profile and lineage potential of ETP have been found to be similar to CLP and LMPP (Luc et al. 2012). Together these data narrow down the competent thymic settling progenitors to subsets of LMPP and/or CLP; although their relative contributions toward T cell development are not known with certainty.

4.2 The Thymic Settling Progenitor Niche

Although the precise identities of the progenitor cells have been elusive, several characteristics of the thymic settling progenitor niche are known. Thymic settling has been suggested to be a "gated" phenomenon. Progenitors settling the thymus are thought to initially occupy niches that only become vacant when these progenitors further differentiate, resulting in periodic waves of thymic entry (Donskoy et al. 2003). The few thymic settling progenitors that have been found have been imaged at the cortico-medullary junction (CMJ) where there are large, postcapillary venules (Lind et al. 2001). This suggests that the thymus settling progenitor niche might be located adjacent to the CMJ (Porritt et al. 2003). Due to the difficulty of studying the cells directly, the number of thymic settling progenitors residing in the niche remains unclear; however, an early estimate was provided by intrathymic injections of HSCs. Approximately 200 HSCs are sufficient to maintain the production of T cells (Spangrude and Scollay 1990). More recently, it was shown that HSC injected into the thymus of ZAP70-deficient mice (which have a block at the CD4/CD8 double positive stage of development) are able to sustain T cell differentiation for considerably longer than physiological thymus settling progenitors (Adjali et al. 2005). Because HSC are more efficient in producing thymocytes than cells that normally settle the thymus, 200 cells is likely to be an underestimate of the size of the niche. Intrathymic transplants of total bone marrow progenitors in ZAP70-deficient mice increased the number of ETP, indicating that this progenitor niche within the thymus is either unsaturated or can be expanded (Vicente et al. 2010). The transplanted cells that engraft the thymus after intrathymic injection of bone marrow progenitors are unable to maintain long-term multilineage hematopoiesis on secondary intravenous transplant (Vicente et al. 2010). Hence, the progenitor niche within the thymus does not support the self-renewal of bonafide HSC.

When fetal thymi are transplanted into mice with competent progenitor populations in bone marrow, resident thymocytes are substituted within four weeks by the progeny of incoming progenitors (Berzins et al. 1998). This observation had suggested that progenitors residing within the thymus were nonrenewing and a constant supply of progenitors imported from the blood was required to maintain progenitors in the thymus, and sustain long-term T cell development. However, recent studies have used thymic transplants into mice lacking the common gamma chain cytokine receptor, or lacking IL-7Rα. In these circumstances, T cell development is sustained on the order of months, much longer than had been previously suggested (Martins et al. 2012; Peaudecerf et al. 2012). These surprising findings indicate that normally, the incoming cells to the progenitor niche force out resident cells based on competition for IL-7 and possibly other growth or differentiation factors such as Flt3 and Notch ligands. In the absence of competition for the niche, progenitors in the thymus can renew the early thymocyte populations for a prolonged period. These data alter our understanding of self-renewal of progenitor populations. Self-renewal ability may not be unique to stem cells, but may depend on whether progenitor cells can compete for relevant niches.

5 Fetal Versus Adult Thymic Settling

Hematopoiesis in the mouse embryo differs from that in the adult. Primitive hematopoietic cells first appear in the yolk sac, but these early progenitors do not generate lymphocytes. Early lymphopoiesis appears to arise independently from HSCs (Yokota et al. 2006). Such "spontaneous" lymphopoiesis occurs at embryonic day 8.5; HSCs with lymphoid potential do not emerge from the aorta-gonad-mesonephros (AGM) region until embryonic day 10.5 (Muller et al. 1994; Li et al. 2012). The main source of hematopoiesis during the remainder of fetal life is the liver (Christensen et al. 2004).

A thymus-parathyroid primordium forms out of the third pharyngeal pouch at embryonic day 9.5; however, the thymic rudiment does not receive an influx of progenitors until embryonic day 10.5 (Hilfer and Brown 1984). Lymphoid progenitors are recruited to the area surrounding the thymic rudiment and are encapsulated within the thymus by approximately embryonic day 12.5 (Jotereau et al. 1987). In the fetal liver, there are well-characterized progenitors that are restricted to T/NK/DC lineages that do not possess B or myeloid cell potential (Kawamoto et al. 2000). These cells express paired immunoglobulin receptor (PIR) and can also be found in the fetal blood and thymus (Ikawa et al. 2004; Masuda et al. 2005). The early fetal T cell progenitors express high levels of Kit and IL-7R and also appear to be restricted to the T/NK/DC lineage (Masuda et al. 2005). This suggests that commitment to the T/NK/DC lineage may occur prior to migration to the thymus in fetal mice. A prethymic T lineage committed population has also been found in adult mice (Dejbakhsh-Jones and Strober 1999; Garcia-Ojeda et al. 2005; Krueger and von Boehmer 2007). It is unclear to what extent T lineage restricted progenitors contribute to thymopoiesis; progenitors settling the thymus may be a heterogeneous population consisting of both lineage-restricted and multipotent cells (Rodewald et al. 1994; Bhandoola et al. 2007).

Thymic homing in the fetus is initiated prior to the vascularization of the thymus primordium. Therefore thymic settling has both prevascular and postvascular stages. During the prevascular phase, chemotaxis is the key component regulating thymic settling. As in the adult, chemokine receptors CCR7 and CCR9 are important to this process (Liu et al. 2006). In addition, CXCR4 has been found to have redundant functions with CCR7 and CCR9 in thymic homing to the prevascular thymus (Calderon and Boehm 2011). In the absence of CCR9, CCR7, and CXCR4, fetal liver progenitors had a 100-fold decrease in thymic homing, whereas any of the double deficient progenitors only had up to 10-fold decrease in thymic homing (Calderon and Boehm 2011). After the fetal thymus vascularizes, thymic homing is thought to occur primarily through the circulation (Dunon et al. 1999). Homing to the vascularized fetal thymus remains dependent on the same three chemokine receptors but the requirement for other molecules must appear so progenitors can migrate in the presence of blood flow (Calderon and Boehm 2011). The progenitors require adhesion molecules such as selectins and integrins to slow the cells, to allow for transendothelial migration into the thymus. The trafficking molecules involved in homing to the vascularized fetal thymus are likely to be similar to the molecules involved in the adult thymus; however, this has not yet been confirmed.

Ephrins (Ephs) are molecules other than chemokines that have been shown to be involved in thymic homing in the fetus (Stimamiglio et al. 2010). Ephs are receptor tyrosine kinases which are essential for pattern formation during embryonic development (Pasquale 2008). EphB2 is expressed on bone marrow progenitors and progenitors deficient in EphB2 exhibit decreased homing to the fetal thymus when in competition with WT cells (Stimamiglio et al. 2010). Furthermore, thymi with TEC deficient in EphB2 do not support normal colonization of progenitors, suggesting a non-cell autonomous effect of EphB2 signaling in TEC on progenitors. In the adult thymus, Ephs are important in many aspects of T cell development; however, more work is needed to determine whether they are also involved in homing (Munoz et al. 2011).

There are several differences in thymopoiesis between the fetus and adult (Fig. 2). In fetal mice, there are much higher levels of progenitors in the blood (Rodewald et al. 1994). Fetal ETP constitute a larger percentage of thymocytes (Hozumi et al. 2008; Belyaev et al. 2012). In adults, fewer progenitors are present in the blood and the ratio of ETP to total thymocytes is lower. However, adult progenitors appear more than 100-fold more efficient at expanding and giving rise to T cells when directly compared by intrathymic injection [(Lu et al. 2005); see Fig. 2]. These differences suggest that cells that settle the thymus are functionally distinct between fetal and adult stages. In fetal mice, more progenitors may settle the thymus to become ETP, however these progenitors have considerably less proliferative capacity than cells with similar phenotypes in adult thymus. Perhaps consistently, gene expression analysis of fetal ETP suggests that they are more

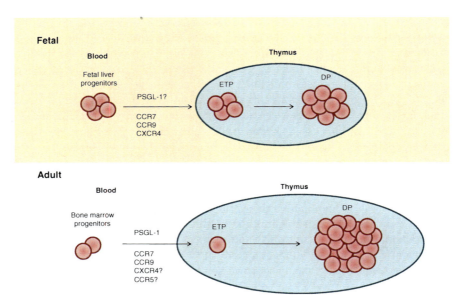

Fig. 2 Differences in T lymphopoiesis between late fetal and adult mice. Progenitors with T cell potential are more abundant in fetal than adult blood. Fetal early thymic progenitors (ETP) constitute a greater fraction of thymocytes and make fewer double positive cells (DP) on a per cell basis

similar to adult DN2 cells than adult ETP (Belyaev et al. 2012). Together, these data suggest that in the adult, increased proliferative ability can compensate for reduced thymic settling.

6 Evolution of the Thymus and Homing

Among jawed vertebrates, elements of progenitor homing to the thymus are conserved. Studies of thymic homing in zebrafish and medaka have shown that similar to mammals, homing relies on chemotaxis (Li et al. 2007; Kissa et al. 2008; Hess and Boehm 2012). The zebrafish genome includes over 100 chemokines as a result of gene duplication during evolution. Two zebrafish chemokines corresponding to mammalian CCL21 and CCL25 were recently characterized (Lu et al. 2012). CCL25 is predominantly expressed in the thymus in both embryos and adult zebrafish, suggesting that CCL25 might play a similar role in homing in zebrafish as in mammals. It is unclear whether CCL21 is also involved in thymic homing in zebrafish as its expression in the pharynx is scattered. It has been suggested that CXCL12 recruits progenitors to the area near the thymic rudiment where they encounter a CCL25 gradient, which then recruits the progenitors into the thymus. Consistently, CCR9 and CXCR4 are implicated in thymic homing in medaka, but CCR7 does not appear to be involved. In mice, CCR7 function in thymic homing is redundant with CCR9 and CXCR4, suggesting that CCR7 was co-opted for thymic homing more recently (Calderon and Boehm 2011). Studies in zebrafish also indicate that hematopoietic progenitors do not require blood flow to home to the thymus and progenitor cells are able to migrate to the thymus by relying primarily on these chemokines (Hess and Boehm 2012).

The most phylogenetically distant existing jawed vertebrate, the cartilaginous fish, possess MHC, TCR, and immunoglobulin, the genetic hallmarks of our adaptive immune system. Jawless fish (agnathans), such as lamprey and hagfish, do not have these genes, suggesting that the common ancestor of jawed vertebrates developed this adaptive immune system after the divergence from jawless vertebrates (Marchalonis and Schluter 1998). Recently, however, jawless vertebrates have been found to have both T and B-like lymphocytes with diverse antigen receptors that confer adaptive immunity (Bajoghli et al. 2009; Guo et al. 2009). The development of the agnathan adaptive immune system appears to parallel that of jawed vertebrates. The T-like lymphocytes (termed VLR-A$^+$ lymphocytes) are present in an immature state within organs found in the gill baskets of lamprey larvae. These organs, termed thymoids, are structures containing lymphoid cells and epithelial cells that express orthologues of Foxn1 and Delta-like ligand (Bajoghli et al. 2011). The evidence suggests that some structural and genetic patterns were co-opted by both the jawed and jawless vertebrate immune systems. The thymoid and the thymus develop in anatomical sites to support T-like and T lymphocyte development that are distinct from other blood cell types, which imposes the requirement for competent progenitors to migrate to the thymoid/thymus.

This raises an interesting question in the field of thymic homing—why did the thymus develop in the gill region at a particular location distant from the site of hematopoiesis? Perhaps the environment of the pharyngeal epithelium provides a unique aggregation of chemokine ligands with Delta-like ligands (Bajoghli et al. 2009; Calderon and Boehm 2012). Notch signaling drives transcriptional programs that cause progenitors to commit to the T lineage (Maillard et al. 2005; Thompson and Zuniga-Pflucker 2011; Rothenberg 2012). Future work can determine if lamprey progenitor homing to the thymus uses the same trafficking molecules and whether the same Notch-dependent transcriptional networks are present in epithelium of the lamprey (Bajoghli et al. 2009). Orthologues of CCR9 and CXCR4 are expressed by mature VLR-A$^+$ lymphocytes in lampreys; however, it is not known whether they are present on VLR-A-lineage competent progenitor populations (Guo et al. 2009).

7 Regeneration of the T Lineage

Regeneration of the immune system is an area of clinical interest particularly in immune compromised patients as well as the elderly. Aging is characterized by changes in bone marrow and thymus that contribute to functional decline in T cell production (Haynes and Swain 2006). Aged bone marrow hematopoietic progenitors lose T lineage potential (Rossi et al. 2005a; Zediak et al. 2007). The architectural structure of the thymus changes as evidenced by disorganization and reduced numbers of TEC (Flores et al. 1999). All T cell progenitor subsets are present within the thymus, but in lower numbers (Min et al. 2004). Total thymic cellularity is reduced, resulting in decreased thymic output of naïve T cells (den Braber et al. 2012). In healthy individuals, it is not clear whether maintaining continuous thymic output is necessary for immune protection; however, reviving thymic function is clearly important for patients that have received cyto-reductive therapies. (Mackall et al. 1995; Hakim et al. 2005).

One example where T cell regeneration is important is following BMT. BMT patients receive conditioning regimens including chemotherapy and/or radiation therapy to severely deplete host hematopoietic cells. Donor cells eventually reconstitute the patient's hematopoietic cell compartments; however, T cells are usually the last of the hematopoietic cell lineages to recover (Wils et al. 2011). This is the case even in autologous transplants, in which patients receive their own bone marrow (Storek et al. 2004). Immune recovery after BMT is often characterized by a prolonged reduction in total $CD4^+$ T cell numbers that leave patients susceptible to infections, correlating with higher rates of mortality (Small et al. 1999; Berger et al. 2008). Following irradiation conditioning, peripheral T cells can be regenerated in several ways: (1) thymus-dependent development of naïve T cells (2) thymus-independent homeostatic proliferation of donor and residual host mature T cells or (3) naïve T cells developing extrathymically (Miller et al. 1963; Dulude et al. 1997). Thymus-independent T cell regeneration results in skewed and limited TCR diversity, which is associated with impaired immunity (Mackall et al. 1996; Yager et al. 2008; Hsieh et al. 2012). Thymus-dependent T cell output can be measured by high levels of T cell receptor excision circles (TREC), which are by-products of V(D)J recombination (Douek et al. 1998). Higher levels of thymic output before BMT, as determined by TREC levels, can predict the success of T cell reconstitution and patient outcome (Svaldi et al. 2003; Chen et al. 2005). In aged individuals who have impaired thymic function, another possible source of naïve T cells are those that develop outside the thymus. This population of T cells was found to develop in athymic nude mice after irradiation conditioning (Dulude et al. 1997). These extrathymically developing cells have been shown to be functional; however, in response to infections, these cells have limited proliferative ability and are a poor substitute for cells derived from the thymus (Blais et al. 2004; Holland et al. 2012).

Several components of thymus-dependent T cell regeneration have been examined following BMT: (1) production of thymic settling bone marrow progenitors, (2) trafficking of the progenitors into the blood and thymus, and (3) the ability of intrathymic environment to support T cell development. Each of these components of T cell development has been found to contribute to the slow recovery of the T cell compartment following BMT. The supply of lymphoid progenitors in the bone marrow is diminished, indicating that the pool of potential thymic settling progenitors is reduced (Zlotoff et al. 2011). Trafficking of progenitors into the thymus is a limiting aspect of T cell recovery after BMT and increasing the number of T cell precursors can enhance T cell reconstitution (Zakrzewski et al. 2006; Dallas et al. 2007; Zlotoff et al. 2011). For BMT patients, the defects in both production of progenitors and trafficking to the thymus can be remedied by increasing the number of T cell competent progenitors that reach the thymus. The intrathymic environment is altered post-BMT in part due to conditioning regimens that deplete the thymus of cortical and medullary TEC (Chung et al. 2001; Williams et al. 2009). Once T cell progenitors settle the thymus, their expansion can be improved by growth factors and cytokines. Keratinocyte growth factor (KGF) protects thymic epithelial cells from injury resulting from the conditioning regimens that precede bone marrow transplantation (Min et al. 2002; Kelly et al. 2008). In both mice and rhesus macaque models of BMT, administration of KGF was shown to improve thymopoiesis and naïve T cell recovery (Alpdogan et al. 2006). In addition, the cytokine IL-22 has recently been shown to be important for TEC regeneration following BMT (Dudakov et al. 2012). Other cytokines have been shown to affect lymphocyte proliferation and survival after BMT. Providing lymphocytes with IL-7 and Flt3 ligand can enhance both prethymic and intrathymic reconstitution. (Fry et al. 2004; Wils et al. 2007; Kenins et al. 2008). Together, these data suggest that enhancing both lymphoid and non-lymphoid tissues can improve regenerative efforts.

The aged thymus can also be regenerated though the ablation of sex steroids; however, the increase in thymic size and naïve T cell output is only temporary (Olsen et al. 1991; Weinberg et al. 2001; Roden et al. 2004). During sex steroid ablation (SSA)-induced regeneration, thymic homing is improved through increased TEC production of CCL25 (Williams et al. 2008). Presently it is unclear whether the genes that dictate normal thymic development and growth are the same as those that control regeneration after SSA. Analysis of the regenerated thymus after SSA suggests that most gene pathways are more similar to those of an old thymus rather than a young thymus (Griffith et al. 2011). Gene expression analysis of the young and regenerated thymus would suggest that the pathways governing initial thymic development and those that dictate regrowth may involve two separate mechanisms. An alternate interpretation of these data is that there are a small number of as yet unidentified shared pathways in both young and regenerated thymi, which dictate thymic size and functionality. The latter possibility suggests that patients who require improved thymic regeneration may benefit from the induction of these pathways.

8 Conclusions

Unlike other blood cells, T cells develop in a sequestered organ distant from the bone marrow. This presents T cell precursors with the additional difficulty of trafficking out of the bone marrow, through the blood, and entering the thymus. Many bone marrow progenitors have the potential to become T cells, yet only a few subsets have the necessary molecules to home to the thymus. These subsets are thought to be within the more lymphoid biased LMPP and CLP populations.

Although the identity of the thymic settling progenitors is becoming clearer, the logic underlying the restriction of thymic homing ability to specific subsets of progenitor cells is not fully understood. Progenitors with greater capacity to produce other blood lineages given unrestricted entry to the thymus might give rise to progeny that compete with early T cell precursors. In the adult, very low numbers of thymic settling progenitors indicate that T lymphopoiesis is mainly driven by proliferative expansion. Why do so few progenitors settle the thymus in adult mice? The small number of progenitors that home to the thymus seems paradoxical as the redundancy of chemokines makes homing robust. The absence of a single chemokine or even two chemokines does not abrogate thymic homing entirely. In zebrafish and medaka, multiple chemokine receptors regulate thymic homing (CCR9 and CXCR4) and more receptors have been implicated in the mouse (CCR7, CCR9, and CXCR4). This redundancy suggests that either many different types of progenitors could home to the thymus each via a separate chemokine receptor or that one specific subset of progenitors can use any of several different chemokine receptors to home to the thymus. As the pool of potential thymic settling progenitors is further narrowed, these two possibilities might be differentiated.

Understanding the identity and properties of the thymic settling cells may contribute to clinical treatments. For example, treating BMT patients with efficient thymic settling progenitors in conjunction with treatments to improve intrathymic reconstitution may help prevent delays in T lineage reconstitution.

References

Adjali O, Vicente RR, Ferrand C, Jacquet C, Mongellaz C, Tiberghien P, Chebli K, Zimmermann VS, Taylor N (2005) Intrathymic administration of hematopoietic progenitor cells enhances T cell reconstitution in ZAP-70 severe combined immunodeficiency. Proc Natl Acad Sci U S A 102:13586–13591

Adolfsson J, Borge OJ, Bryder D, Theilgaard-Monch K, Astrand-Grundstrom I, Sitnicka E, Sasaki Y, Jacobsen SE (2001) Upregulation of Flt3 Expression within the bone marrow Lin(-)Sca1(+)c- kit(+) stem cell compartment is accompanied by loss of self-renewal capacity. Immunity 15:659–669

Adolfsson J, Mansson R, Buza-Vidas N, Hultquist A, Liuba K, Jensen CT, Bryder D, Yang L, Borge OJ, Thoren LA, Anderson K, Sitnicka E, Sasaki Y, Sigvardsson M, Jacobsen SE (2005) Identification of Flt3 + lympho-myeloid stem cells lacking erythro-megakaryocytic potential a revised road map for adult blood lineage commitment. Cell 121:295–306

Allman D, Sambandam A, Kim S, Miller JP, Pagan A, Well D, Meraz A, Bhandoola A (2003) Thymopoiesis independent of common lymphoid progenitors. Nat Immunol 4:168–174

Alpdogan O, Hubbard VM, Smith OM, Patel N, Lu S, Goldberg GL, Gray DH, Feinman J, Kochman AA, Eng JM, Suh D, Muriglan SJ, Boyd RL, van den Brink MR (2006) Keratinocyte growth factor (KGF) is required for postnatal thymic regeneration. Blood 107:2453–2460

Trafficking to the Thymus

Anderson G, Jenkinson EJ, Moore NC, Owen JJ (1993) MHC class II-positive epithelium and mesenchyme cells are both required for T-cell development in the thymus. Nature 362:70–73

Bajoghli B, Aghaallaei N, Hess I, Rode I, Netuschil N, Tay BH, Venkatesh B, Yu JK, Kaltenbach SL, Holland ND, Diekhoff D, Happe C, Schorpp M, Boehm T (2009) Evolution of genetic networks underlying the emergence of thymopoiesis in vertebrates. Cell 138:186–197

Bajoghli B, Guo P, Aghaallaei N, Hirano M, Strohmeier C, McCurley N, Bockman DE, Schorpp M, Cooper MD, Boehm T (2011) A thymus candidate in lampreys. Nature 470:90–94

Bell JJ, Bhandoola A (2008) The earliest thymic progenitors for T cells possess myeloid lineage potential. Nature 452:764–767

Belyaev NN, Biro J, Athanasakis D, Fernandez-Reyes D, Potocnik AJ (2012) Global transcriptional analysis of primitive thymocytes reveals accelerated dynamics of T cell specification in fetal stages. Immunogenetics 64:591–604

Benz C, Martins VC, Radtke F, Bleul CC (2008) The stream of precursors that colonizes the thymus proceeds selectively through the early T lineage precursor stage of T cell development. J Exp Med 205:1187–1199

Berger M, Figari O, Bruno B, Raiola A, Dominietto A, Fiorone M, Podesta M, Tedone E, Pozzi S, Fagioli F, Madon E, Bacigalupo A (2008) Lymphocyte subsets recovery following allogeneic bone marrow transplantation (BMT): CD4 + cell count and transplant-related mortality. Bone Marrow Transplant 41:55–62

Berzins SP, Boyd RL, Miller JF (1998) The role of the thymus and recent thymic migrants in the maintenance of the adult peripheral lymphocyte pool. J Exp Med 187:1839–1848

Bhandoola A, von Boehmer H, Petrie HT, Zuniga-Pflucker JC (2007) Commitment and developmental potential of extrathymic and intrathymic T cell precursors: plenty to choose from. Immunity 26:678–689

Blais ME, Gerard G, Martinic MM, Roy-Proulx G, Zinkernagel RM, Perreault C (2004) Do thymically and strictly extrathymically developing T cells generate similar immune responses? Blood 103:3102–3110

Boehm T, Bleul CC (2006) Thymus-homing precursors and the thymic microenvironment. Trends Immunol 27:477–484

Boehm T, Bleul CC, Schorpp M (2003) Genetic dissection of thymus development in mouse and zebrafish. Immunol Rev 195:15–27

Broxmeyer HE, Orschell CM, Clapp DW, Hangoc G, Cooper S, Plett PA, Liles WC, Li X, Graham-Evans B, Campbell TB, Calandra G, Bridger G, Dale DC, Srour EF (2005) Rapid mobilization of murine and human hematopoietic stem and progenitor cells with AMD3100, a CXCR4 antagonist. J Exp Med 201:1307–1318

Calderon L, Boehm T (2011) Three chemokine receptors cooperatively regulate homing of hematopoietic progenitors to the embryonic mouse thymus. Proc Natl Acad Sci U S A 108:7517–7522

Calderon L, Boehm T (2012) Synergistic, context-dependent, and hierarchical functions of epithelial components in thymic microenvironments. Cell 149:159–172

Chen X, Barfield R, Benaim E, Leung W, Knowles J, Lawrence D, Otto M, Shurtleff SA, Neale GA, Behm FG, Turner V, Handgretinger R (2005) Prediction of T-cell reconstitution by assessment of T-cell receptor excision circle before allogeneic hematopoietic stem cell transplantation in pediatric patients. Blood 105:886–893

Chi AW, Chavez A, Xu L, Weber BN, Shestova O, Schaffer A, Wertheim G, Pear WS, Izon D, Bhandoola A (2010) Identification of Flt3(+)CD150(-) myeloid progenitors in adult mouse bone marrow that harbor T lymphoid developmental potential. Blood 118:2723–2732

Christensen JL, Weissman IL (2001) Flk-2 is a marker in hematopoietic stem cell differentiation: a simple method to isolate long-term stem cells. Proc Natl Acad Sci U S A 98:14541–14546

Christensen JL, Wright DE, Wagers AJ, Weissman IL (2004) Circulation and chemotaxis of fetal hematopoietic stem cells. PLoS Biol 2:E75

Chung B, Barbara-Burnham L, Barsky L, Weinberg K (2001) Radiosensitivity of thymic interleukin-7 production and thymopoiesis after bone marrow transplantation. Blood 98:1601–1606

Ciofani M, Zuniga-Pflucker JC (2007) The thymus as an inductive site for T lymphopoiesis. Annu Rev Cell Dev Biol 23:463–493

Dallas MH, Varnum-Finney B, Martin PJ, Bernstein ID (2007) Enhanced T-cell reconstitution by hematopoietic progenitors expanded ex vivo using the Notch ligand Delta1. Blood 109:3579–3587

De Obaldia ME, Bell JJ, Bhandoola A (2013) Early T-cell progenitors are the major granulocyte precursors in the adult mouse thymus. Blood 121:64–71

Dejbakhsh-Jones S, Strober S (1999) Identification of an early T cell progenitor for a pathway of T cell maturation in the bone marrow. Proc Natl Acad Sci U S A 96:14493–14498

den Braber I, Mugwagwa T, Vrisekoop N, Westera L, Mögling R, de Boer AB, Willems N, Schrijver EH, Spierenburg G, Gaiser K, Mul E, Otto SA, Ruiter AF, Ackermans MT, Miedema F, Borghans JA, de Boer RJ, Tesselaar K (2012) Maintenance of peripheral naive T cells: a mouse-main divide. Immunity 36:288–297

Donskoy E, Goldschneider I (1992) Thymocytopoiesis is maintained by blood-borne precursors throughout postnatal life. A study in parabiotic mice. J Immunol 148:1604–1612

Donskoy E, Foss D, Goldschneider I (2003) Gated importation of prothymocytes by adult mouse thymus is coordinated with their periodic mobilization from bone marrow. J Immunol 171:3568–3575

Douek DC, McFarland RD, Keiser PH, Gage EA, Massey JM, Haynes BF, Polis MA, Haase AT, Feinberg MB, Sullivan JL, Jamieson BD, Zack JA, Picker LJ, Koup RA (1998) Changes in thymic function with age and during the treatment of HIV infection. Nature 396:690–695

Dudakov JA, Hanash AM, Jenq RR, Young LF, Ghosh A, Singer NV, West ML, Smith OM, Holland AM, Tsai JJ, Boyd RL, van den Brink MR (2012) Interleukin-22 drives endogenous thymic regeneration in mice. Science 336:91–95

Dulude G, Brochu S, Fontaine P, Baron C, Gyger M, Roy DC, Perreault C (1997) Thymic and extrathymic differentiation and expansion of T lymphocytes following bone marrow transplantation in irradiated recipients. Exp Hematol 25:992–1004

Dunon D, Allioli N, Vainio O, Ody C, Imhof BA (1999) Quantification of T-cell progenitors during ontogeny: thymus colonization depends on blood delivery of progenitors. Blood 93:2234–2243

Flores KG, Li J, Sempowski GD, Haynes BF, Hale LP (1999) Analysis of the human thymic perivascular space during aging. J Clin Invest 104:1031–1039

Fry TJ, Sinha M, Milliron M, Chu YW, Kapoor V, Gress RE, Thomas E, Mackall CL (2004) Flt3 ligand enhances thymic-dependent and thymic-independent immune reconstitution. Blood 104:2794–2800

Garcia-Ojeda ME, Dejbakhsh-Jones S, Chatterjea-Matthes D, Mukhopadhyay A, BitMansour A, Weissman IL, Brown JM, Strober S (2005) Stepwise development of committed progenitors in the bone marrow that generate functional T cells in the absence of the thymus. J Immunol 175:4363–4373

Golan K, Vagima Y, Ludin A, Itkin T, Cohen-Gur S, Kalinkovich A, Kollet O, Kim C, Schajnovitz A, Ovadya Y, Lapid K, Shivtiel S, Morris AJ, Ratajczak MZ, Lapidot T (2012) S1P promotes murine progenitor cell egress and mobilization via S1P1-mediated ROS signaling and SDF-1 release. Blood 119:2478–2488

Goodman JW, Hodgson GS (1962) Evidence for stem cells in the peripheral blood of mice. Blood 19:702–714

Gossens K, Naus S, Corbel SY, Lin S, Rossi FM, Kast J, Ziltener HJ (2009) Thymic progenitor homing and lymphocyte homeostasis are linked via S1P-controlled expression of thymic P-selectin/CCL25. J Exp Med 206:761–778

Griffith AV, Fallahi M, Venables T, Petrie HT (2011) Persistent degenerative changes in thymic organ function revealed by an inducible model of organ regrowth. Aging Cell 11:169–177

Guo P, Hirano M, Herrin BR, Li J, Yu C, Sadlonova A, Cooper MD (2009) Dual nature of the adaptive immune system in lampreys. Nature 459:796–801

Hakim FT, Memon SA, Cepeda R, Jones EC, Chow CK, Kasten-Sportes C, Odom J, Vance BA, Christensen BL, Mackall CL, Gress RE (2005) Age-dependent incidence, time course, and consequences of thymic renewal in adults. J Clin Invest 115:930–939

Haynes L, Swain SL (2006) Why aging T cells fail: implications for vaccination. Immunity 24:663–666

Hess I, Boehm T (2012) Intravital imaging of thymopoiesis reveals dynamic lympho-epithelial interactions. Immunity 36:298–309

Hilfer SR, Brown JW (1984) The development of pharyngeal endocrine organs in mouse and chick embryos. Scan Electron Microsc 4:2009–2022

Holland AM, Zakrzewski JL, Tsai JJ, Hanash AM, Dudakov JA, Smith OM, West ML, Singer NV, Brill J, Sun JC, van den Brink MR (2012) Extrathymic development of murine T cells after bone marrow transplantation. J Clin Invest 122:4716–4726

Hozumi K, Mailhos C, Negishi N, Hirano K, Yahata T, Ando K, Zuklys S, Hollander GA, Shima DT, Habu S (2008) Delta-like 4 is indispensable in thymic environment specific for T cell development. J Exp Med 205:2507–2513

Hsieh MY, Hong WH, Lin JJ, Lee WI, Lin KL, Wang HS, Chen SH, Yang CP, Jaing TH, Huang JL (2012) T-cell receptor excision circles and repertoire diversity in children with profound T-cell immunodeficiency. J Microbiol Immunol Infect. http://dx.doi.org/10.1016/j.jmii.2012.06.003

Igarashi H, Gregory SC, Yokota T, Sakaguchi N, Kincade PW (2002) Transcription from the RAG1 locus marks the earliest lymphocyte progenitors in bone marrow. Immunity 17:117–130

Ikawa T, Masuda K, Lu M, Minato N, Katsura Y, Kawamoto H (2004) Identification of the earliest prethymic T-cell progenitors in murine fetal blood. Blood 103:530–537

Inlay MA, Bhattacharya D, Sahoo D, Serwold T, Seita J, Karsunky H, Plevritis SK, Dill DL, Weissman IL (2009) Ly6d marks the earliest stage of B-cell specification and identifies the branchpoint between B-cell and T-cell development. Genes Dev 23:2376–2381

Iwasaki H, Akashi K (2007) Hematopoietic developmental pathways: on cellular basis. Oncogene 26:6687–6696

Izon DJ (2008) T-cell development: thymus-settling progenitors: settled? Immunol Cell Biol 86:552–553

Jin Y, Wu MX (2008) Requirement of Galphai in thymic homing and early T cell development. Mol Immunol 45:3401–3410

Jotereau F, Heuze F, Salomon-Vie V, Gascan H (1987) Cell kinetics in the fetal mouse thymus: precursor cell input, proliferation, and emigration. J Immunol 138:1026–1030

Kawamoto H, Ikawa T, Ohmura K, Fujimoto S, Katsura Y (2000) T cell progenitors emerge earlier than B cell progenitors in the murine fetal liver. Immunity 12:441–450

Kelly RM, Highfill SL, Panoskaltsis-Mortari A, Taylor PA, Boyd RL, Hollander GA, Blazar BR (2008) Keratinocyte growth factor and androgen blockade work in concert to protect against conditioning regimen-induced thymic epithelial damage and enhance T-cell reconstitution after murine bone marrow transplantation. Blood 111:5734–5744

Kenins L, Gill JW, Boyd RL, Hollander GA, Wodnar-Filipowicz A (2008) Intrathymic expression of Flt3 ligand enhances thymic recovery after irradiation. J Exp Med 205:523–531

King AG, Horowitz D, Dillon SB, Levin R, Farese AM, MacVittie TJ, Pelus LM (2001) Rapid mobilization of murine hematopoietic stem cells with enhanced engraftment properties and evaluation of hematopoietic progenitor cell mobilization in rhesus monkeys by a single injection of SB-251353, a specific truncated form of the human CXC chemokine GRObeta. Blood 97:1534–1542

Kissa K, Murayama E, Zapata A, Cortes A, Perret E, Machu C, Herbomel P (2008) Live imaging of emerging hematopoietic stem cells and early thymus colonization. Blood 111:1147–1156

Koch U, Radtke F (2011) Mechanisms of T cell development and transformation. Annu Rev Cell Dev Biol 27:539–562

Kondo M, Weissman IL, Akashi K (1997) Identification of clonogenic common lymphoid progenitors in mouse bone marrow. Cell 91:661–672

Krueger A, von Boehmer H (2007) Identification of a T lineage-committed progenitor in adult blood. Immunity 26:105–116

Krueger A, Garbe AI, von Boehmer H (2006) Phenotypic plasticity of T cell progenitors upon exposure to Notch ligands. J Exp Med 203:1977–1984

Krueger A, Willenzon S, Lyszkiewicz M, Kremmer E, Forster R (2010) CC chemokine receptor 7 and 9 double-deficient hematopoietic progenitors are severely impaired in seeding the adult thymus. Blood 115:1906–1912

Kunisaki Y, Frenette PS (2012) The secrets of the bone marrow niche: Enigmatic niche brings challenge for HSC expansion. Nat Med 18:864–865

Lai AY, Kondo M (2007) Identification of a bone marrow precursor of the earliest thymocytes in adult mouse. Proc Natl Acad Sci U S A 104:6311–6316

Lei Y, Liu C, Saito F, Fukui Y, Takahama Y (2009) Role of DOCK2 and DOCK180 in fetal thymus colonization. Eur J Immunol 39:2695–2702

Levesque JP, Hendy J, Takamatsu Y, Simmons PJ, Bendall LJ (2003) Disruption of the CXCR4/CXCL12 chemotactic interaction during hematopoietic stem cell mobilization induced by GCSF or cyclophosphamide. J Clin Invest 111:187–196

Li F, Wilkins PP, Crawley S, Weinstein J, Cummings RD, McEver RP (1996) Post-translational modifications of recombinant P-selectin glycoprotein ligand-1 required for binding to P- and E-selectin. J Biol Chem 271:3255–3264

Li J, Iwanami N, Hoa VQ, Furutani-Seiki M, Takahama Y (2007) Noninvasive intravital imaging of thymocyte dynamics in medaka. J Immunol 179:1605–1615

Li Z, Lan Y, He W, Chen D, Wang J, Zhou F, Wang Y, Sun H, Chen X, Xu C, Li S, Pang Y, Zhang G, Yang L, Zhu L, Fan M, Shang A, Ju Z, Luo L, Ding Y, Guo W, Yuan W, Yang X, Liu B (2012) Mouse embryonic head as a site for hematopoietic stem cell development. Cell Stem Cell 11:663–675

Lind EF, Prockop SE, Porritt HE, Petrie HT (2001) Mapping precursor movement through the postnatal thymus reveals specific microenvironments supporting defined stages of early lymphoid development. J Exp Med 194:127–134

Liu F, Poursine-Laurent J, Link DC (2000) Expression of the G-CSF receptor on hematopoietic progenitor cells is not required for their mobilization by G-CSF. Blood 95:3025–3031

Liu C, Saito F, Liu Z, Lei Y, Uehara S, Love P, Lipp M, Kondo S, Manley N, Takahama Y (2006) Coordination between CCR7- and CCR9-mediated chemokine signals in prevascular fetal thymus colonization. Blood 108:2531–2539

Lu M, Tayu R, Ikawa T, Masuda K, Matsumoto I, Mugishima H, Kawamoto H, Katsura Y (2005) The earliest thymic progenitors in adults are restricted to T, NK, and dendritic cell lineage and have a potential to form more diverse TCRbeta chains than fetal progenitors. J Immunol 175:5848–5856

Lu IN, Chiang BL, Lou KL, Huang PT, Yao CC, Wang JS, Lin LD, Jeng JH, Chang BE (2012) Cloning, expression and characterization of CCL21 and CCL25 chemokines in zebrafish. Dev Comp Immunol 38:203–214

Luc S, Luis TC, Boukarabila H, Macaulay IC, Buza-Vidas N, Bouriez-Jones T, Lutteropp M, Woll PS, Loughran SJ, Mead AJ, Hultquist A, Brown J, Mizukami T, Matsuoka S, Ferry H, Anderson K, Duarte S, Atkinson D, Soneji S, Domanski A, Farley A, Sanjuan-Pla A, Carella C, Patient R, de Bruijn M, Enver T, Nerlov C, Blackburn C, Godin I, Jacobsen SE (2012) The earliest thymic T cell progenitors sustain B cell and myeloid lineage potential. Nat Immunol 13:412–419

Lymperi S, Ferraro F, Scadden DT (2010) The HSC niche concept has turned 31. Has our knowledge matured? Ann N Y Acad Sci 1192:12–18

Mackall CL, Fleisher TA, Brown MR, Andrich MP, Chen CC, Feuerstein IM, Horowitz ME, Magrath IT, Shad AT, Steinberg SM et al (1995) Age, thymopoiesis, and CD4 + T-lymphocyte regeneration after intensive chemotherapy. N Engl J Med 332:143–149

Mackall CL, Bare CV, Granger LA, Sharrow SO, Titus JA, Gress RE (1996) Thymic-independent T cell regeneration occurs via antigen-driven expansion of peripheral T cells resulting in a repertoire that is limited in diversity and prone to skewing. J Immunol 156:4609–4616

Maillard I, Fang T, Pear WS (2005) Regulation of lymphoid development, differentiation, and function by the Notch pathway. Annu Rev Immunol 23:945–974

Mansson R, Zandi S, Welinder E, Tsapogas P, Sakaguchi N, Bryder D, Sigvardsson M (2011) Single-cell analysis of the common lymphoid progenitor compartment reveals functional and molecular heterogeneity. Blood 115:2601–2609

Marchalonis JJ, Schluter SF (1998) A stochastic model for the rapid emergence of specific vertebrate immunity incorporating horizontal transfer of systems enabling duplication and combinational diversification. J Theor Biol 193:429–444

Marshall E, Woolford LB, Lord BI (1997) Continuous infusion of macrophage inflammatory protein MIP-1alpha enhances leucocyte recovery and haemopoietic progenitor cell mobilization after cyclophosphamide. Br J Cancer 75:1715–1720

Martin CH, Aifantis I, Scimone ML, von Andrian UH, Reizis B, von Boehmer H, Gounari F (2003) Efficient thymic immigration of B220+lymphoid-restricted bone marrow cells with T precursor potential. Nat Immunol 4:866–873

Martins VC, Ruggiero E, Schlenner SM, Madan V, Schmidt M, Fink PJ, von Kalle C, Rodewald HR (2012) Thymus-autonomous T cell development in the absence of progenitor import. J Exp Med 209:1409–1417

Masuda K, Kubagawa H, Ikawa T, Chen CC, Kakugawa K, Hattori M, Kageyama R, Cooper MD, Minato N, Katsura Y, Kawamoto H (2005) Prethymic T-cell development defined by the expression of paired immunoglobulin-like receptors. EMBO J 24:4052–4060

Mendez-Ferrer S, Michurina TV, Ferraro F, Mazloom AR, Macarthur BD, Lira SA, Scadden DT, Ma'ayan A, Enikolopov GN, Frenette PS (2010) Mesenchymal and haematopoietic stem cells form a unique bone marrow niche. Nature 466:829–834

Mikkola HK, Orkin SH (2006) The journey of developing hematopoietic stem cells. Development 133:3733–3744

Miller JP, Doak SMA, Cross AM (1963) Role of the Thymus in Recovery of the Immune Mechanism in the Irradiated Adult Mouse. Exp Biol Med 112:785–792

Min D, Taylor PA, Panoskaltsis-Mortari A, Chung B, Danilenko DM, Farrell C, Lacey DL, Blazar BR, Weinberg KI (2002) Protection from thymic epithelial cell injury by keratinocyte growth factor: a new approach to improve thymic and peripheral T-cell reconstitution after bone marrow transplantation. Blood 99:4592–4600

Min H, Montecino-Rodriguez E, Dorshkind K (2004) Reduction in the developmental potential of intrathymic T cell progenitors with age. J Immunol 173:245–250

Misslitz A, Pabst O, Hintzen G, Ohl L, Kremmer E, Petrie HT, Forster R (2004) Thymic T cell development and progenitor localization depend on CCR7. J Exp Med 200:481–491

Morris GP, Allen PM (2012) How the TCR balances sensitivity and specificity for the recognition of self and pathogens. Nat Immunol 13:121–128

Morrison SJ, Wandycz AM, Hemmati HD, Wright DE, Weissman IL (1997) Identification of a lineage of multipotent hematopoietic progenitors. Development 124:1929–1939

Muller AM, Medvinsky A, Strouboulis J, Grosveld F, Dzierzak E (1994) Development of hematopoietic stem cell activity in the mouse embryo. Immunity 1:291–301

Munoz JJ, Cejalvo T, Alonso-Colmenar LM, Alfaro D, Garcia-Ceca J, Zapata A (2011) Eph/Ephrin-mediated interactions in the thymus. NeuroImmunoModulation 18:271–280

Murray LJ, Luens KM, Estrada MF, Bruno E, Hoffman R, Cohen RL, Ashby MA, Vadhan-Raj S (1998) Thrombopoietin mobilizes CD34+cell subsets into peripheral blood and expands multilineage progenitors in bone marrow of cancer patients with normal hematopoiesis. Exp Hematol 26:207–216

Olsen NJ, Watson MB, Henderson GS, Kovacs WJ (1991) Androgen deprivation induces phenotypic and functional changes in the thymus of adult male mice. Endocrinology 129:2471–2476

Osawa M, Hanada K, Hamada H, Nakauchi H (1996) Long-term lymphohematopoietic reconstitution by a single CD34-low/negative hematopoietic stem cell. Science 273:242–245

Pasquale EB (2008) Eph-ephrin bidirectional signaling in physiology and disease. Cell 133:38–52

Peaudecerf L, Lemos S, Galgano A, Krenn G, Vasseur F, Di Santo JP, Ezine S, Rocha B (2012) Thymocytes may persist and differentiate without any input from bone marrow progenitors. J Exp Med 209:1401–1408

Pelus LM, Bian H, King AG, Fukuda S (2004) Neutrophil-derived MMP-9 mediates synergistic mobilization of hematopoietic stem and progenitor cells by the combination of G-CSF and the chemokines GRObeta/CXCL2 and GRObetaT/CXCL2delta4. Blood 103:110–119

Perry SS, Welner RS, Kouro T, Kincade PW, Sun XH (2006) Primitive lymphoid progenitors in bone marrow with T lineage reconstituting potential. J Immunol 177:2880–2887

Petit I, Szyper-Kravitz M, Nagler A, Lahav M, Peled A, Habler L, Ponomaryov T, Taichman RS, Arenzana-Seisdedos F, Fujii N, Sandbank J, Zipori D, Lapidot T (2002) G-CSF induces stem cell mobilization by decreasing bone marrow SDF-1 and up-regulating CXCR4. Nat Immunol 3:687–694

Pettengell R, Woll PJ, Chang J, Coutinho L, Testa NG, Crowther D (1994) Effects of erythropoietin on mobilisation of haemopoietic progenitor cells. Bone Marrow Transplant 14:125–130

Pitchford SC, Furze RC, Jones CP, Wengner AM, Rankin SM (2009) Differential mobilization of subsets of progenitor cells from the bone marrow. Cell Stem Cell 4:62–72

Porritt HE, Gordon K, Petrie HT (2003) Kinetics of steady-state differentiation and mapping of intrathymic-signaling environments by stem cell transplantation in nonirradiated mice. J Exp Med 198:957–962

Porritt HE, Rumfelt LL, Tabrizifard S, Schmitt TM, Zuniga-Pflucker JC, Petrie HT (2004) Heterogeneity among DN1 prothymocytes reveals multiple progenitors with different capacities to generate T cell and non-T cell lineages. Immunity 20:735–745

Pruijt JF, Verzaal P, van Os R, de Kruijf EJ, van Schie ML, Mantovani A, Vecchi A, Lindley IJ, Willemze R, Starckx S, Opdenakker G, Fibbe WE (2002) Neutrophils are indispensable for hematopoietic stem cell mobilization induced by interleukin-8 in mice. Proc Natl Acad Sci U S A 99:6228–6233

Ratajczak MZ, Lee H, Wysoczynski M, Wan W, Marlicz W, Laughlin MJ, Kucia M, Janowska-Wieczorek A, Ratajczak J (2010) Novel insight into stem cell mobilization-plasma sphingosine-1-phosphate is a major chemoattractant that directs the egress of hematopoietic stem progenitor cells from the bone marrow and its level in peripheral blood increases during mobilization due to activation of complement cascade/membrane attack complex. Leukemia 24:976–985

Robertson P, Means TK, Luster AD, Scadden DT (2006) CXCR4 and CCR5 mediate homing of primitive bone marrow-derived hematopoietic cells to the postnatal thymus. Exp Hematol 34:308–319

Roden AC, Moser MT, Tri SD, Mercader M, Kuntz SM, Dong H, Hurwitz AA, McKean DJ, Celis E, Leibovich BC, Allison JP, Kwon ED (2004) Augmentation of T cell levels and responses induced by androgen deprivation. J Immunol 173:6098–6108

Rodewald HR, Kretzschmar K, Takeda S, Hohl C, Dessing M (1994) Identification of pro-thymocytes in murine fetal blood: T lineage commitment can precede thymus colonization. EMBO J 13:4229–4240

Rossi DJ, Bryder D, Zahn JM, Ahlenius H, Sonu R, Wagers AJ, Weissman IL (2005a) Cell intrinsic alterations underlie hematopoietic stem cell aging. Proc Natl Acad Sci U S A 102:9194–9199

Rossi FM, Corbel SY, Merzaban JS, Carlow DA, Gossens K, Duenas J, So L, Yi L, Ziltener HJ (2005b) Recruitment of adult thymic progenitors is regulated by P-selectin and its ligand PSGL-1. Nat Immunol 6:626–634

Rothenberg EV (2012) Transcriptional drivers of the T-cell lineage program. Curr Opin Immunol 24:132–138

Ruiz P, Wiles MV, Imhof BA (1995) Alpha 6 integrins participate in pro-T cell homing to the thymus. Eur J Immunol 25:2034–2041

Schwarz BA, Bhandoola A (2004) Circulating hematopoietic progenitors with T lineage potential. Nat Immunol 5:953–960

Schwarz BA, Sambandam A, Maillard I, Harman BC, Love PE, Bhandoola A (2007) Selective thymus settling regulated by cytokine and chemokine receptors. J Immunol 178:2008–2017

Scimone ML, Aifantis I, Apostolou I, von Boehmer H, von Andrian UH (2006) A multistep adhesion cascade for lymphoid progenitor cell homing to the thymus. Proc Natl Acad Sci U S A 103:7006–7011

Serwold T, Ehrlich LI, Weissman IL (2009) Reductive isolation from bone marrow and blood implicates common lymphoid progenitors as the major source of thymopoiesis. Blood 113:807–815

Shortman K, Wu L (1996) Early T lymphocyte progenitors. Annu Rev Immunol 14:29–47

Simmons PJ, Masinovsky B, Longenecker BM, Berenson R, Torok-Storb B, Gallatin WM (1992) Vascular cell adhesion molecule-1 expressed by bone marrow stromal cells mediates the binding of hematopoietic progenitor cells. Blood 80:388–395

Small TN, Papadopoulos EB, Boulad F, Black P, Castro-Malaspina H, Childs BH, Collins N, Gillio A, George D, Jakubowski A, Heller G, Fazzari M, Kernan N, MacKinnon S, Szabolcs P, Young JW, O'Reilly RJ (1999) Comparison of immune reconstitution after unrelated and related T-cell-depleted bone marrow transplantation: effect of patient age and donor leukocyte infusions. Blood 93:467–480

Spangrude GJ, Scollay R (1990) Differentiation of hematopoietic stem cells in irradiated mouse thymic lobes. Kinetics and phenotype of progeny. J Immunol 145:3661–3668

Spangrude GJ, Heimfeld S, Weissman IL (1988) Purification and characterization of mouse hematopoietic stem cells. Science 241:58–62

Springer TA (1994) Traffic signals for lymphocyte recirculation and leukocyte emigration: the multistep paradigm. Cell 76:301–314

Stimamiglio MA, Jimenez E, Silva-Barbosa SD, Alfaro D, Garcia-Ceca JJ, Munoz JJ, Cejalvo T, Savino W, Zapata A (2010) EphB2-mediated interactions are essential for proper migration of T cell progenitors during fetal thymus colonization. J Leukoc Biol 88:483–494

Storek J, Douek DC, Keesey JC, Boehmer L, Storer B, Maloney DG (2003) Low T cell receptor excision circle levels in patients thymectomized 25–54 years ago. Immunol Lett 89:91–92

Storek J, Staver JH, Porter BA, Maloney DG (2004) The thymus is typically small at 1 year after autologous or allogeneic T-cell-replete hematopoietic cell transplantation into adults. Bone Marrow Transplant 34:829–830

Stritesky GL, Jameson SC, Hogquist KA (2012) Selection of self-reactive T cells in the thymus. Annu Rev Immunol 30:95–114

Sultana DA, Zhang SL, Todd SP, Bhandoola A (2012) Expression of functional P-selectin glycoprotein ligand 1 on hematopoietic progenitors is developmentally regulated. J Immunol 188:4385–4393

Svaldi M, Lanthaler AJ, Dugas M, Lohse P, Pescosta N, Straka C, Mitterer M (2003) T-cell receptor excision circles: a novel prognostic parameter for the outcome of transplantation in multiple myeloma patients. Br J Haematol 122:795–801

Thompson PK, Zuniga-Pflucker JC (2011) On becoming a T cell, a convergence of factors kick it up a Notch along the way. Semin Immunol 23:350–359

Uehara S, Grinberg A, Farber JM, Love PE (2002) A role for CCR9 in T lymphocyte development and migration. J Immunol 168:2811–2819

Ueno T, Saito F, Gray DH, Kuse S, Hieshima K, Nakano H, Kakiuchi T, Lipp M, Boyd RL, Takahama Y (2004) CCR7 signals are essential for cortex-medulla migration of developing thymocytes. J Exp Med 200:493–505

van Den Brink M, Leen AM, Baird K, Merchant M, Mackall C, Bollard CM (2013) Enhancing Immune Reconstitution: from Bench to Bedside. Biol Blood Marrow Transplant 19:S79-S83

Vicari AP, Figueroa DJ, Hedrick JA, Foster JS, Singh KP, Menon S, Copeland NG, Gilbert DJ, Jenkins NA, Bacon KB, Zlotnik A (1997) TECK: a novel CC chemokine specifically expressed by thymic dendritic cells and potentially involved in T cell development. Immunity 7:291–301

Vicente R, Adjali O, Jacquet C, Zimmermann VS, Taylor N (2010) Intrathymic transplantation of bone marrow-derived progenitors provides long-term thymopoiesis. Blood 115:1913–1920

Wallis VJ, Leuchars E, Chwalinski S, Davies AJ (1975) On the sparse seeding of bone marrow and thymus in radiation chimaeras. Transplantation 19:2–11

Weinberg K, Blazar BR, Wagner JE, Agura E, Hill BJ, Smogorzewska M, Koup RA, Betts MR, Collins RH, Douek DC (2001) Factors affecting thymic function after allogeneic hematopoietic stem cell transplantation. Blood 97:1458–1466

Weinreich MA, Hogquist KA (2008) Thymic emigration: when and how T cells leave home. J Immunol 181:2265–2270

Williams KM, Lucas PJ, Bare CV, Wang J, Chu YW, Tayler E, Kapoor V, Gress RE (2008) CCL25 increases thymopoiesis after androgen withdrawal. Blood 112:3255–3263

Williams KM, Mella H, Lucas PJ, Williams JA, Telford W, Gress RE (2009) Single cell analysis of complex thymus stromal cell populations: rapid thymic epithelia preparation characterizes radiation injury. Clin Transl Sci 2:279–285

Willimann K, Legler DF, Loetscher M, Roos RS, Delgado MB, Clark-Lewis I, Baggiolini M, Moser B (1998) The chemokine SLC is expressed in T cell areas of lymph nodes and mucosal lymphoid tissues and attracts activated T cells via CCR7. Eur J Immunol 28:2025–2034

Wils EJ, Braakman E, Verjans GM, Rombouts EJ, Broers AE, Niesters HG, Wagemaker G, Staal FJ, Lowenberg B, Spits H, Cornelissen JJ (2007) Flt3 ligand expands lymphoid progenitors prior to recovery of thymopoiesis and accelerates T cell reconstitution after bone marrow transplantation. J Immunol 178:3551–3557

Wils EJ, van der Holt B, Broers AE, Posthumus-van Sluijs SJ, Gratama JW, Braakman E, Cornelissen JJ (2011) Insufficient recovery of thymopoiesis predicts for opportunistic infections in allogeneic hematopoietic stem cell transplant recipients. Haematologica 96:1846–1854

Wilson A, Laurenti E, Oser G, van der Wath RC, Blanco-Bose W, Jaworski M, Offner S, Dunant CF, Eshkind L, Bockamp E, Lio P, Macdonald HR, Trumpp A (2008) Hematopoietic stem cells reversibly switch from dormancy to self-renewal during homeostasis and repair. Cell 135:1118–1129

Winkler IG, Pettit AR, Raggatt LJ, Jacobsen RN, Forristal CE, Barbier V, Nowlan B, Cisterne A, Bendall LJ, Sims NA, Levesque JP (2012) Hematopoietic stem cell mobilizing agents G-CSF, cyclophosphamide or AMD3100 have distinct mechanisms of action on bone marrow HSC niches and bone formation. Leukemia 26:1594–1601

Wright DE, Wagers AJ, Gulati AP, Johnson FL, Weissman IL (2001) Physiological migration of hematopoietic stem and progenitor cells. Science 294:1933–1936

Wurbel MA, Philippe JM, Nguyen C, Victorero G, Freeman T, Wooding P, Miazek A, Mattei MG, Malissen M, Jordan BR, Malissen B, Carrier A, Naquet P (2000) The chemokine TECK is expressed by thymic and intestinal epithelial cells and attracts double- and single-positive thymocytes expressing the TECK receptor CCR9. Eur J Immunol 30:262–271

Yager EJ, Ahmed M, Lanzer K, Randall TD, Woodland DL, Blackman MA (2008) Age-associated decline in T cell repertoire diversity leads to holes in the repertoire and impaired immunity to influenza virus. J Exp Med 205:711–723

Yokota T, Huang J, Tavian M, Nagai Y, Hirose J, Zuniga-Pflucker JC, Peault B, Kincade PW (2006) Tracing the first waves of lymphopoiesis in mice. Development 133:2041–2051

Zakrzewski JL, Kochman AA, Lu SX, Terwey TH, Kim TD, Hubbard VM, Muriglan SJ, Suh D, Smith OM, Grubin J, Patel N, Chow A, Cabrera-Perez J, Radhakrishnan R, Diab A, Perales MA, Rizzuto G, Menet E, Pamer EG, Heller G, Zuniga-Pflucker JC, Alpdogan O, van den Brink MR (2006) Adoptive transfer of T-cell precursors enhances T-cell reconstitution after allogeneic hematopoietic stem cell transplantation. Nat Med 12:1039–1047

Zediak VP, Maillard I, Bhandoola A (2007) Multiple prethymic defects underlie age-related loss of T progenitor competence. Blood 110:1161–1167

Zlotoff DA, Bhandoola A (2011) Hematopoietic progenitor migration to the adult thymus. Ann N Y Acad Sci 1217:122–138

Zlotoff DA, Sambandam A, Logan TD, Bell JJ, Schwarz BA, Bhandoola A (2010) CCR7 and CCR9 together recruit hematopoietic progenitors to the adult thymus. Blood 115:1897–1905

Zlotoff DA, Zhang SL, De Obaldia ME, Hess PR, Todd SP, Logan TD, Bhandoola A (2011) Delivery of progenitors to the thymus limits T-lineage reconstitution after bone marrow transplantation. Blood 118:1962–1970

The CD4/CD8 Lineages: Central Decisions and Peripheral Modifications for T Lymphocytes

Hirokazu Tanaka and Ichiro Taniuchi

Abstract CD4$^+$ helper and CD8$^+$ cytotoxic T cells, two major subsets of $\alpha\beta$TCR expressing lymphocytes, are differentiated from common precursor CD4$^+$CD8$^+$ double-positive (DP) thymocytes. Bifurcation of the CD4$^+$/CD8$^+$ lineages in the thymus is a multilayered process and is thought to culminate in a loss of developmental plasticity between these functional subsets. Advances in the last decade have deepened our understanding of the transcription control mechanisms governing CD4 versus CD8 lineage commitment. Reciprocal expression and antagonistic interplay between two transcription factors, ThPOK and Runx3, is crucial for driving thymocyte decisions between these two cell fates. Here, we first focus on the regulation of ThPOK expression and its role in directing helper T cell development. We then discuss a novel aspect of the ThPOK/Runx3 axis in modifying CD4$^+$ T cell function upon exposure to gut microenvironment.

Contents

1 Introduction.. 114
2 A Central Role for ThPOK in Transcription Factor Networks That
  Guide Helper T Cell Development.. 115
3 *Thpok* Repression Is Required for CD8 T Cell Development 119
4 Activation of Genes Required for CD8$^+$ T Cell Development 121
5 CD4/CD8 Lineage Modification in the Gut: Unappreciated Novel
  Plasticity in CD4$^+$ T Cell... 123
6 Concluding Remarks 125
References.. 126

H. Tanaka · I. Taniuchi (✉)
Laboratory for Transcriptional Regulation,
RIKEN Research Center for Allergy and Immunology,
1-7-22 Suehiro-cho, Tsurumi-ku, Yokohama,
Kanagawa 230-0045, Japan
e-mail: taniuchi@rcai.riken.jp

Current Topics in Microbiology and Immunology (2014) 373: 113–129
DOI: 10.1007/82_2013_323
© Springer-Verlag Berlin Heidelberg 2013
Published Online: 24 April 2013

1 Introduction

Understanding the mechanisms that underlie cell fate determination of multipotent precursors is an overarching goal of developmental biology. Maturation of the mammalian immune system requires many rounds of fate determination at developmental branch points, resulting in a diverse array of immune soldiers armed with specific functions (Carpenter and Bosselut 2010). T lymphopoiesis has long been recognized as a useful model for lineage-decision mechanisms, in large part because a plethora of cell surface markers are available for the high-resolution separation of distinct cellular subsets. For instance, developmental stages of thymocytes are readily identified by expression patterns of the CD4 and CD8 co-receptors (Ellmeier et al. 1999; Singer and Bosselut 2004). Following their migration to the thymus, early thymocytes progenitors (ETP) lack surface CD4 and CD8, and therefore are referred to as CD4$^-$CD8$^-$ double-negative (DN) thymocytes. Once the DN cells have assembled a functional *Tcrb* gene and express the corresponding protein in the form of a pre-TCR (β-selection), they proliferate and activate expression of the *Cd4* and *Cd8* genes, generating a large CD4$^+$CD8$^+$ double-positive (DP) thymocyte population. During the DN to DP transition, thymocytes assemble *Tcra*, culminating in the expression of a mature $\alpha\beta$TCR on the surface of DP thymocytes. To further differentiate into mature thymocytes, DP cells must pass another selection process, known as a positive/negative selection, during which reactivity of $\alpha\beta$TCR to self-peptide/MHC is evaluated (Germain 2002). Current evidence indicates that only a small portion of DP thymocytes survive positive/negative selection and face a choice between two alternative fates; to become helper or cytotoxic T cells.

A major factor involved in the cytotoxic/helper T lineage decision is the type of MHC molecules engaged by DP thymocyte precursors during selection. While MHC class I-selected thymocytes mature into cytotoxic T cells, acquiring a CD4$^-$CD8$^+$ single-positive (CD8 SP) surface phenotype, those selected by MHC class II differentiate into the helper T lineage and become CD4$^+$CD8$^-$ single-positive (CD4 SP) thymocytes (Koller et al. 1990; Grusby et al. 1991). Thus, in addition to TCR-MHC specificities, expression of CD4/CD8 co-receptors perfectly matches the segregation of helper/cytotoxic functional lineages. Accordingly, this lineage selection process is often referred to as CD4/CD8 lineage choice. The links between TCR/MHC specificity, co-receptor expression, and cell fate decisions suggest that differences in TCR signal quality, quantity, or duration are converted into distinct gene regulatory networks (Gascoigne and Palmer 2011). The output of these networks then confers distinct cellular function and co-receptor expression during the terminal stages of thymocyte maturation.

Together, these unique, well-defined characteristics have promoted helper/cytotoxic T lineage decisions to a favored position among models to understand how bi-potential precursors choose particular cell fates following their exposure to differentiation cues. As such, the mechanisms that regulate helper/cytotoxic-lineage choices have been studied extensively and, as a consequence, several models have been proposed and challenged (Singer et al. 2008). Although the molecular nature of nuclear sensors that convert TCR signals into developmental programs remains obscure, advances made in the last decade have identified nuclear factors involved in the CD4/CD8 lineage choice. In this review, we will mainly focus on two transcription factors, ThPOK and Runx3, as the major drivers of CD4$^+$ helper and CD8$^+$ cytotoxic differentiation, respectively. We will discuss how expression of these two factors is regulated and how, in turn, ThPOK and Runx3 regulate developmental programs leading to the helper or cytotoxic T lineages. Lastly, we discuss a new aspect of extrathymic modifications to helper T cell function that occur specifically upon antigen stimulation in the gut environment, and highlight the roles of ThPOK/Runx3 in reshaping a functional identity of $\alpha\beta$T cells (Fig. 1).

2 A Central Role for ThPOK in Transcription Factor Networks That Guide Helper T Cell Development

ThPOK, also known as Zbtb7b or cKrox, is encoded by the *Zbtb7b/Thpok* gene (hereafter referred to as *Thpok*) and belongs to a BTB (Broad complex, tramtrack, bric-a-brac) domain containing zinc-finger family of transcription factors. Genetic studies in mice demonstrated that ThPOK is a central transcription factor for CD4$^+$ T cell development. Initial evidence came from Kappes and colleagues, who identified a genetic locus responsible for the helper-deficient (HD) phenotype of a spontaneous mouse mutant (Dave et al. 1998). Using a straightforward positional cloning approach, they found a single nucleotide change in the *Thpok* gene, which is causative for the HD phenotype and introduces a missense mutation (R389G) at the second zinc-finger domain of the ThPOK protein (He et al. 2005). In HD mice, almost all MHC class II-selected thymocytes are redirected into the alternative CD4$^-$CD8$^+$ cytotoxic T cell lineage, a phenotype shared with ThPOK-deficient mice generated by gene targeting. Thus, the "*hd*" mutation likely abrogates ThPOK function. More importantly, ectopic expression of ThPOK at the DP stage onward not only rescues the HD phenotype but also redirects all MHC class I-selected thymocytes to differentiate into CD4$^+$ T cells (He et al. 2005; Sun et al. 2005). These genetic data clearly demonstrate that ThPOK expression is not only essential, but also sufficient, to guide post-selection thymocytes into the CD4$^+$CD8$^-$ lineage and acquire the functional properties of helper T cells.

Other studies have addressed the role of ThPOK in maintaining commitment of peripheral CD4$^+$ cells to the helper T lineage. Upon transfer into immunodeficient recipients, CD4$^+$ T cells from which *Thpok* had been conditionally excised upregulate genes that are characteristic of the cytotoxic T lineage, including *Cd8* and *Granzyme B (GzmB)* (Wang et al. 2008a). This finding revealed that continuous extrathymic expression of ThPOK is necessary to maintain helper-lineage gene signatures, which are established during differentiation in the thymus. Moreover, helper-lineage commitment may be dependent on ThPOK dosage. In

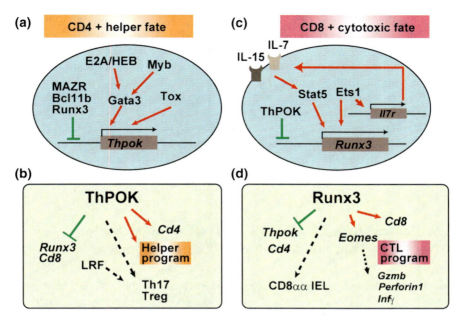

Fig. 1 Roles of ThPOK and Runx3 in transcriptional networks for functional bifurcation of CD4-helper and CD8-cytotoxic lineages. a *Thpok* gene expression is repressed in preselection thymocytes and cytotoxic-lineage cells via the *Thpok* silencer, whose activity depends on functions of Runx, MAZR, and Bcl11b transcription factors. Gata3, Tox, c-Myb transcription factors are involved in not only *Thpok* activation but also in executing commitment to CD4-helper lineage. b Role of ThPOK in programming toward CD4+ helper T cells. ThPOK antagonizes the *Cd4* silencer activity for CD4 expression in helper lineage cells and represses cytotoxic lineage signature genes such as *Runx3* and *Cd8*. Roles of ThPOK in regulating Treg and Th17 differentiation in lamina propria are in part overlapped with a closely related transcription factor, LRF. c Cytokine signaling mediated by Stat5, a target of IL-7 in thymus, promotes *Runx3* expression. Ets1 transcription factor is also necessary for appropriate *Runx3* expression. Ets1 also activate the gene encoding IL-7 receptor. On the other hand, *Runx3* expression is repressed by ThPOK in CD4+ helper T cells. d Runx3 is necessary not only for repression of helper-lineage signature genes, such as *Thpok* and *Cd4*, but also for activation of cytotoxic-lineage marker gene, *Cd8*. Runx3 plays central roles in activating genes specific for cytotoxic T cells such as *Gzmb*, *perforin1*, *Infγ*, and *eomes*. Of note, Runx3 is also necessary for development of CD8αα IELs. Red lines with arrows and green lines indicate activating and repressive effects on transcription, respectively. Dashed lines indicate promotion of differentiation

mice harboring a hypomorphic *Thpok* allele, which produces reduced levels of ThPOK protein due to the insertion of a neomycin resistant gene, MHC class II-selected cells are partially redirected into the CD4−CD8+ lineage (Egawa and Littman 2008; Wang et al. 2008b). Importantly, CD4+CD8− T cells generated in these mutant mice fail to repress cytotoxic-related genes, such as *Runx3* and *GzmB*, while helper-lineage signatures, such as CD40L induction, were impaired as well. Consistent with these observations, when the developmental potential of CD4+ thymocytes expressing ThPOK at intermediate or high levels was tested in vitro, CD4$^-$CD8$^+$ T cells emerged specifically from the cells expressing intermediate amounts of ThPOK (Muroi et al. 2008). Thus, simple induction of ThPOK expression is insufficient for full commitment to the helper lineage. Rather, full commitment likely consists of sequential steps, including escalation of ThPOK levels.

The ThPOK dosage-dependence during helper-lineage commitment has spurred numerous studies of *Thpok* gene regulation. One genetic study showed that a kinetic increase in ThPOK levels is regulated at the transcriptional level via the activity of a proximal enhancer (PE), which locates ~ 1 kb downstream from exon Ib in the *Thpok* gene (Muroi et al. 2008). Removal of the PE from a *Thpok*^{gfp} reporter allele significantly attenuates GFP expression, mainly at the later stages of differentiation in MHC class II-selected cells. Failure to upregulate ThPOK in the PE-deficient thymocytes leads to a 50 % drop in the number of MHC class II-selected cells that differentiate into the CD4$^+$CD8$^-$ phenotyped cells. Another recent study found that the ThPOK protein is acetylated on at least three lysine residues by the p300 acetyltransferase (Zhang et al. 2010). This post-translational modification may stabilize the ThPOK protein by curtailing its ubiquitin-mediated degradation (Zhang et al. 2010). Thus, both transcriptional and post-translational mechanisms play a key role in setting ThPOK protein levels.

Despite these advances, the mechanisms by which ThPOK activates the developmental program conferring a helper-lineage expression signature remain less characterized. Recent studies indicate that ThPOK requires additional input from other transcription factors in programming the helper-lineage commitment. In this regard, the Gata-3 transcription factor, known to play a key role at multiple stages of T cell development (Hosoya et al. 2010), is also essential to direct differentiation toward the helper lineage. Inactivation of *Gata3* at the DP stage inhibits generation of the CD4-SP, but not the CD8-SP, thymocyte subset (Zhu et al. 2006). Indeed, Gata-3 may be an upstream activator of the *Thpok* gene based on the lack of *Thpok* expression in post-selection Gata3-deficient thymocytes and the presence of Gata-3 bindings at two regions in the *Thpok* gene (Wang et al. 2008b). However, *Thpok* gene regulation is not the only role of Gata-3 in CD4$^+$ lineage commitment. Transgenic complementation of ThPOK in the Gata-3 deficient cells fail to rescue differentiation of MHC class II-selected thymocytes into CD4$^+$ T cells, whereas this ectopic ThPOK expression could redirect MHC class I-selected thymocytes toward CD4$^+$ T cells (Wang et al. 2008b). Thus, in addition to its function as an upstream factor for *Thpok* gene activation, Gata-3 has another essential, yet uncharacterized, role(s) in CD4$^+$ T cell development.

Similar to Gata-3, the *Tox* gene, which encodes a high-mobility group (HMG) box protein, is essential for both CD4$^+$ T cell development and *Thpok* activation (Aliahmad and Kaye 2008). Since Tox deficiency does not alter *Gata-3* expression levels, the Tox- and Gata3-mediated pathways may function independently to initiate *Thpok* expression. Transgenic expression of ThPOK in *Tox*$^{-/-}$ mice restores CD4$^+$ T cell development to some extent; however, the "rescued" CD4$^+$ T cells are compromised in establishing an appropriate helper-lineage gene signature (Aliahmad et al. 2011). This finding indicates that TOX-dependent functions, beyond those involved in *Thpok* gene activation, are also necessary for proper helper T cell development.

The E protein transcription factors, HEB and E2A, are also critical for helper-lineage development. If both HEB and E2A activities are removed from DP thymocytes, differentiation of CD8 SP, but not CD4 SP, thymocytes are induced independent of TCR signaling (Jones and Zhuang 2007). Conversely, deletion of two E protein inhibitors, Id2 and Id3, which would result in enhanced E protein activities, drive CD4 SP, but not CD8 SP, thymocyte differentiation (Jones-Mason et al. 2012). To further address the mechanisms of how E proteins regulate CD4-lineage development, Gata-3 expression during thymocyte differentiation was examined in either HEB/E2A or Id2/Id3 doubly deficient mice. While Gata-3 induction was impaired in the former genotype, it was enhanced in mice lacking the Id inhibitors (Jones-Mason et al. 2012). These results demonstrate that HEB and E2A proteins may work as upstream activators to enhance Gata-3 expression in post-selection thymocytes. Taken together, current evidence indicates that several ThPOK-independent regulatory mechanisms operate in parallel with ThPOK-dependent pathways to direct MHC class II-restricted thymocytes into the CD4$^+$ T cell lineage.

Several lines of evidence suggest that, in addition to the dominant ThPOK-driven pathway, other partially redundant pathways may also contribute to helper T cell differentiation (Carpenter et al. 2012). Although the "HD" phenotype caused by the *hd* mutation in *Thpok* stands for "HD" (Dave et al. 1998), a small number of CD4$^+$ T cells, including both CD4$^+$CD8$^-$ and CD4$^+$CD8$^+$ subsets, are present in the peripheral T cell pools of these animals. Thus, ThPOK deficiency is "leaky" since redirection of MHC class II-selected cells into the CD4$^-$CD8$^+$ cytotoxic T lineage is incomplete. Moreover, activation of the redirected MHC class II-restricted CD4$^-$CD8$^+$ cells from ThPOK-deficient mice induced both CD4 and CD40L expression, two hallmarks of the helper T cell lineage, to some extent (Carpenter et al. 2012). These findings show that a subset of helper-related signatures is latently retained in the redirected MHC class II-selected cells from ThPOK-deficient mice, a conclusion supported by the retention of active epigenetic modifications at the *Cd4* and *Cd40 l* loci.

Emerging study suggests that LRF, also known as Pokemon, is one factor that may have at least some functional overlap with ThPOK during thymocyte differentiation (Carpenter et al. 2012). LRF is encoded by the *Zbtb7a* gene and plays a critical role in lymphoid lineage commitment, especially for differentiation of B-lymphocytes in the bone marrow (Maeda et al. 2007). LRF functions, in part, via the regulation of Notch-ligand (Delta-like 4) expression on erythroblasts (Lee et al. 2012). Amino acid sequence comparisons indicate that LRF is the closest homologue to ThPOK among mammalian BTB-POZ family members (Lee and Maeda 2012), further supporting the possibility of functional redundancy between these factors. Phenotypic analyses of ThPOK/LRF double-mutant mice have provided genetic evidence that the two factors have at least some functional overlap in regulating thymocyte differentiation. Specifically, ThPOK/LRF double-deficient precursors could not give rise to CD4$^+$CD8$^-$ or CD4$^+$ CD8$^+$ cells (Carpenter et al. 2012). In addition, the ThPOK/LRF double mutants lack a subset of $CD4^+CD8^+$ T cells found in the lamina propria of ThPOK-deficient animals, which are positive for $Foxp3^+$ or $Ror\gamma t^+$ expression, markers for Treg or Th17 effectors cells, respectively. Thus, LRF overlaps partially with ThPOK function to support Treg and Th17 differentiation. However, LRF does not provide compensatory functions for ThPOK in all developmental pathways. For example, ThPOK-deficient mice completely lack IL4-producing Th2 cells in their lamina propria, indicating a unique function for this protein in supporting Th2 cell differentiation that cannot be rescued by LRF. Instead, Th2 cell differentiation of ThPOK-deficient cells can be restored in vitro by perturbation of Runx3 function (Carpenter et al. 2012), suggesting that ThPOK preserves Th2 potency via its unique ability to repress Runx3 expression. This finding is consistent in part with the restored development of $CD4^+$ T cells in ThPOK-deficient mice following elimination of Runx function in the thymus. Specifically, DP thymocytes lacking both ThPOK and $Cbf\beta$ protein, an essential partner for Runx family proteins (Wang et al. 1996), primarily give rise to $CD4^+CD8^-$ T cells (Egawa and Littman 2008). However, it should be noted that the requirement for ThPOK-mediated repression of Runx3 in directing the differentiation of IL4-producing Th2 cells could reflect a dominant Runx3 function in silencing the *Il4* gene. A similar mechanism was shown to act during Th1 differentiation through Runx3 binding to an *Il4* gene silencer element (Djuretic et al. 2007; Naoe et al. 2007).

3 *Thpok* Repression Is Required for CD8 T Cell Development

It has become apparent that programming of helper-lineage T cell development is regulated by transcription factors in both ThPOK-dependent and independent manners. There also exists at least one factor, LRF, which is partially redundant with ThPOK functions. However, ectopic expression of ThPOK alone is sufficient to redirect MHC class I-selected thymocytes into the $CD4^+$ T cells, demonstrating that ThPOK occupies the central position in transcription factor networks that govern helper T cell development. A corollary of this conclusion is that mechanisms must exist to suppress *Thpok* expression in cells destined for the cytotoxic lineage. Accordingly, studies focused on unraveling *Thpok* gene regulation have moved to the forefront for understanding transcriptional programs involved in CD4/CD8 lineage decisions. Using different approaches, two groups led by Kappes and Taniuchi independently reported that an active repressive mechanism ensures helper-lineage-specific expression of *Thpok*. While Kappes' group performed functional characterization of DNase hypersensitive sites (He et al. 2008), Taniuchi's group performed phenotypic analyses of Runx mutant mice combined with ChIP to identify Runx binding sites in the *Thpok* locus (Setoguchi et al. 2008). These studies uncovered a transcriptional silencer within the distal regulatory element (DRE) located 3.2 kb upstream from the *Thpok* P1 promoter. Importantly, this silencer element, hereafter referred to as *Thpok* silencer, is essential for *Thpok* repression during the differentiation of MHC class I-selected cells (He et al. 2008; Setoguchi et al. 2008). Deletion of the core *Thpok* silencer in mice dramatically impairs the production of CD8$^+$ cells, confirming the functional relevance of silencer-mediated *Thpok* regulation for cytotoxic T cell differentiation *in vivo* (Setoguchi et al. 2008).

Although the two studies identified similar silencer regions, the role of Runx transcription factors in the enforcement of silencer activity has been controversial (Hedrick 2008). Although depletion of the *trans*-acting Runx/Cbfβ complex de-represses *Thpok* in MHC class I-selected cells (Setoguchi et al. 2008), a substantial silencer activity remains in mutant versions of the silencer that lacks Runx sites when measured using a transgenic reporter assay (He et al. 2008). More recently, we have generated mice harboring specific mutations at two Runx sites in the endogenous *Thpok* silencer and find a nearly abrogation of its silencer activity for *Thpok* repression in MHC class I-selected cells (Tanaka et al. 2013). These data confirm a key role for Runx proteins in *Thpok* silencer function, which may establish a platform for the recruitment of repressive nuclear complexes to the DRE region. One remaining puzzle is that Runx complexes associate with the *Thpok* silencer even in cells expressing the *Thpok* gene (Setoguchi et al. 2008). As such, Runx binding is essential, but not sufficient, for silencer function, and the "on–off" switch for silencer activity may rely on additional layers of regulation.

These findings underscore the importance of other unidentified factors in *Thpok* silencer function. In this regard, phenotypic analyses of mouse strains with mutations in genes encoding the Mazr and Bcl11b revealed a requirement for these transcription factors in *Thpok* repression. MAZR is another member of the BTB-POZ family that had been established previously as a negative regulator of *Cd8* gene expression (Bilic et al. 2006). Subsequent analyses of *Mazr*-deficient mice revealed that a small proportion of MHC class I-selected cells are redirected into the CD4$^+$ lineage (Sakaguchi et al. 2010). Furthermore, partial de-repression of *Thpok* was observed in MHC class I-selected cells lacking MAZR as measured by a *Thpok-gfp* reporter allele. The transcription factor Bcl11b has been shown to regulate T cell development at multiple levels. In mice lacking Bcl11b, early thymocyte progenitors fail to differentiate beyond the DN2 stage, where full commitment to the T-lineage occurs (Ikawa et al. 2010; Li et al. 2010a, b). Moreover, positive selection at the DP to SP transition is severely impaired in the Bcl11b mutants (Albu et al. 2007). Gene expression profiling revealed a premature activation of *Thpok* in DP thymocytes that are deficient for Bcl11b (Kastner et al. 2010). Importantly, both Bcl11b and MAZR proteins bind the DRE in the *Thpok* gene as measured by ChIP (Kastner et al. 2010; Sakaguchi et al. 2010). Together, these findings suggest that Runx, MAZR, and Bcl11b function cooperatively to regulate *Thpok* silencer activity.

4 Activation of Genes Required for CD8$^+$ T Cell Development

During differentiation of MHC class I-selected cells into CD8$^+$ cytotoxic cells, expression of the *Cd8* locus, which consists of the *Cd8α* and *Cd8β* genes, is dynamically regulated. The *Cd8* gene is temporally repressed after positive selection and then reactivated specifically in MHC class I-selected cells, a process designated as "coreceptor reversal" (Brugnera et al. 2000). However, the molecular mechanisms that underlie dynamic expression of *Cd8* during the CD4/CD8 lineage decision remain obscure. Previous studies identified at least five enhancers in the murine *Cd8* locus, each of which exhibits distinct stage-specific activity in reporter transgene assays (Ellmeier et al. 1997, 1998; Hostert et al. 1998). Among those *Cd8* enhancers, the E8I enhancer is important for *Cd8* expression at later stages of thymocyte development. When the E8I is juxtaposed with the *Cd8a* promoter, it drives transgene expression specifically in CD8$^+$ SP thymocytes, CD8$^+$ T cells, and CD8$αα^+$ intraepithelial lymphocytes (IEL), but not in DP thymocytes (Ellmeier et al. 1997). Of note, deletion of the E8I in mice did not significantly impact CD8 expression in resting cytotoxic-lineage cells, indicating the presence of additional enhancer that compensates E8I function. However, E8I-deficient cytotoxic T cells fail to maintain CD8 expression at normal levels following activation, suggesting a greater importance for supporting *Cd8* expression during an active immune response (Ellmeier et al. 1997; Hassan et al. 2011). Although the relevant enhancer(s) that compensate for loss of the E8I in resting cytotoxic T cells has not been identified, *Cd8* enhancer redundancy has been reported. Combined deletions of E8II with E8III synergistically reduce CD8 expression on DP cells, albeit in a variegated manner (Ellmeier et al. 1998). Collectively, these findings demonstrate that expression of the murine *Cd8* gene during T cell development is regulated in an ordered and coordinated manner via the action of multiple enhancers (Ellmeier et al. 1998; Hostert et al. 1998).

Several transcription factors that bind to and regulate the activities of *Cd8* enhancers have been identified. The binding of Runx/Cbf$β$ complexes to *Cd8* enhancers was first reported by Sato et al. (2005) and more recent studies have confirmed the functional contribution of this transcription factor complex to *Cd8* activation. Activated CD8$^+$ T cells deficient for Runx3 or Cbf$β$ fail to maintain CD8 expression at normal levels (Hassan et al. 2011). In addition, mice lacking a distal promoter-derived Runx3 protein, a major Runx3 variant expressed in the CD8-lineage cells (Egawa et al. 2007), lack CD8$αα$ IELs (Pobezinsky et al. 2012). Thus, current evidence indicates that Runx3/Cbf$β$ complexes are essential components of protein complexes that bind and activate *Cd8* enhancers, in particular the E8I. Runx3 also associates with genes encoding Granzyme B, Perforin, and IFN-$γ$ in effector CD8$^+$ CTLs (Cruz-Guilloty et al. 2009). Indeed, CD8$^+$ T cells lacking Runx3 fail to efficiently induce this set of genes as well as the gene encoding Eomesodermin (Eomes), a central transcription factor in the differentiation program of effector CD8$^+$ T cells. As such, Runx3 plays a central role in activating cytotoxic-related signatures genes.

In addition to Runx3, the transcription factor SATB1, special AT-rich binding protein 1, is reported to be a positive regulator of *Cd8* gene expression. Inactivation of the *Satb1* gene during early thymocyte differentiation impairs CD8-lineage cell development (Alvarez et al. 2000). Recently, Bcl11b was also shown to associate with *Cd8* enhancers (Vanvalkenburgh et al. 2011). Conditional deletion of Bcl11b at later stages of thymocyte maturation attenuates CD8 expression, cytotoxic activity, and clonal expansion of CD8-lineage cells (Vanvalkenburgh et al. 2011). Finally, as discussed above, MAZR negatively regulates *Cd8* expression at the DN to DP transition through its binding to certain *Cd8* enhancers (Bilic et al. 2006).

Based on the accumulated body of data, we can conclude that thymocytes and mature cytotoxic T cells must strike a delicate balance between positive and negative factors for proper regulation of the *Cd8* gene expression. Since repression of ThPOK is critical for differentiation of CD8$^+$ cytotoxic T cells, one could speculate that a central function of ThPOK is to antagonize transcriptional programs that confer cytotoxic gene expression signatures, including activation of the *Cd8* gene. Recent ChIP studies confirm that the *Cd8* locus is, in fact, a target of ThPOK antagonism, which binds to multiple regions in the *Cd8 gene*, including the E8I (Rui et al. 2012; Mucida et al. 2013). With regard to its repressive function, ThPOK associates with histone deacetylase (HDAC) family proteins (Rui et al. 2012). These observations suggest an antagonistic action by ThPOK for *Cd8* repression; ThPOK induces histone deacetylation at the *Cd8* locus via its competitive binding to enhancers, in turn excluding factors that activate enhancer function. At the functional level, retroviral transduction of ThPOK expression vectors into mature CD8$^+$ T cells, full committed CD8-lineage cells, still reduces *Cd8* expression albeit less efficiently (Jenkinson et al. 2007). On the contrary, the ThPOK-*Cd8* connection is less clear in preselection DP thymocytes. In one study, a ThPOK transgene significantly reduced *Cd8* transcription in DP thymocytes (Rui et al. 2012), whereas surface CD8 levels seemed to be unaffected by ectopic ThPOK expression in another study of DP thymocytes (He et al. 2005). Although this matter must be clarified, it remains likely that the mechanisms by which ThPOK represses *Cd8* differ at distinct stages of T cell development.

Given its potential role in the regulation of cytotoxic-lineage genes, an important goal is to understand how Runx3 expression is controlled during T cell development. As mentioned above, ThPOK represses the *Runx3* gene during helper-lineage commitment in the thymus, although few mechanistic insights into this process are available. The factors that activate *Runx3* gene expression are beginning to emerge. Stat5, which acts downstream of IL-7 signaling, promotes *Runx3* expression after cessation of TCR signaling in post-selection thymocytes (Park et al. 2010). In what may be a related finding, the Ets1 transcription factor contributes to the activation of *Runx3* gene expression (Zamisch et al. 2009) and helps to maintain IL-7 receptor expression (Grenningloh et al. 2011). To test the role of cytokine signaling in setting transcriptional programs for cytotoxic-lineage commitment, the common receptor γ chain was conditionally deleted in thymocytes. This loss-of-function approach showed that IL-7 and IL-15 signals are crucial for differentiation of CD8 SP thymocytes. However, Runx3 expression was unexpectedly induced to some extent in the absence of γc cytokine signaling (McCaughtry et al. 2012). A clue to this apparent paradox is provided by experiments showing that overexpression of SOCS1, an inhibitor of multiple cytokines signals, inhibits *Runx3* expression. Thus, it remains possible that signals from non-γc cytokine receptor also contribute to the positive regulation of *Runx3* expression in the CD8 lineage. Alternatively, γc cytokine signals may not be required for *Runx*3 activation. Since the available information about *cis*-regulatory regions in the *Runx3* locus is scant, future studies will have to unravel the mechanisms of *Runx3* induction in developing thymocytes.

Notwithstanding these uncertainties, compelling evidence supports an antagonistic interplay between ThPOK and Runx3 in cross-regulating the expression of their corresponding genes. This negative feedback strategy for gene regulation is a crucial component of the mechanisms governing helper versus cytotoxic fate decisions (Egawa and Taniuchi 2009). In addition, Runx3 and ThPOK serve as counterbalances to regulate target genes that functionally define the two distinct lineages, including the genes that encode for CD4 and CD8 co-receptors. This type of transcription factor antagonism appears to be a common theme in mechanisms that regulate cell fate determination (Singh 2007). As such, the antagonistic ThPOK/Runx3 axis and its central role in helper/cytotoxic-lineage decisions may reflect a common evolutionary strategy for the emergence of new developmental branch points from ancestral progenitors.

5 CD4/CD8 Lineage Modification in the Gut: Unappreciated Novel Plasticity in CD4⁺ T Cell

Conventional wisdom has long held that commitment to the CD4⁺ helper or CD8⁺ cytotoxic T cell subset in the thymus confers specific cellular functions that are maintained stably in the periphery. Recent progress in elucidating epigenetic mechanisms that control gene expression status has provided important new information on how cell identity is established and maintained along complex developmental pathways. In the field of T cell development, studies of *Cd4* gene regulation in CD8⁺ cytotoxic T cells have generated new paradigms for understanding epigenetic mechanisms that maintain repression. Although the intronic *Cd4* silencer is essential to establish the initial repressive state at *Cd4* during CD8-lineage commitment in the thymus, repression is maintained after removal of the silencer from mature CD8⁺ T cells in the periphery (Zhu et al. 2004). This pioneering work demonstrated that epigenetic mechanisms are engaged in peripheral cells to stably maintain *Cd4* silencing, whereas initial repression in the thymus is mediated genetically via the silencer element. Runx3 binding to the *Cd4* silencer was shown to be an essential component of the mechanisms that establish stable epigenetic silencing of the *Cd4* locus (Taniuchi et al. 2002; Woolf et al. 2003). This finding raises the possibility that Runx3 is also involved in sealing the repressive fate of other helper-related genes, thereby conferring stability to cytotoxic T cell identity. Given that the *Thpok* silencer depends on Runx binding (Setoguchi et al. 2008), we recently demonstrated that epigenetic mechanisms of gene silencing are also employed at the *Thpok* locus. Similar to stable silencing of *Cd4*, once *Thpok* repression has been established particularly at the proximal promoter, it is then maintained in mature $CD8^+$ T cells after removal of the *Thpok* silencer (Tanaka et al. 2013).

Contrary to our increasing knowledge of *Cd4* and *Thpok* repression, the epigenetic mechanisms that control stable repression of *Cd8* during helper-lineage commitment remain unclear. For instance, a negative *cis*-regulatory element(s) is yet to be identified for the *Cd8* locus, if one even exists. However, as discussed above, several lines of evidence have revealed parallel roles for ThPOK and Runx3 in the repression of *Cd8* and *Cd4* genes, respectively (Taniuchi et al. 2002; Rui et al. 2012). Inactivation of *Thpok* in mature $CD4^+$ T cells, using Cre/loxP mediated recombination, results in inappropriate *Cd8* reactivation (Wang et al. 2008a), suggesting that the repressive state of *Cd8* gene in $CD4^+$ T cells is reversible. Likewise, *in vitro* stimulation of $CD4^+$ T cells with a combination of α-CD3 antibody and cytokine TGF-β diminishes ThPOK levels and induces CD8 expression, albeit to a limited extent (Konkel et al. 2011). However, the physiologic relevance of ThPOK down-regulation and consequential CD8 reexpression remained to be established in an *in vivo* setting.

One recent study has now provided clear evidence for the physiological relevance of this potential form of lineage plasticity. Upon transfer of naïve $CD4^+CD8^-$ T cells containing the *Thpok-gfp* reporter allele into lympho-deficient RAG KO recipient, a proportion of these cells not only down-regulated Thpok-gfp but also reexpressed $CD8\alpha$ in both the small and large intestine, but not in other peripheral lymphoid tissues such as spleen and lymph nodes (Mucida et al. 2013). This "reawakening" of *Cd8* was confirmed by *in vivo* fate mapping, which showed that $CD4^+CD8\alpha\alpha^+$ IEL differentiate from CD4-lienage cells under normal, immune-sufficient conditions (Mucida et al. 2013). However, similar to other IEL subsets, the absence of $CD4^+CD8\alpha\alpha^+$ IEL in germ-free mice indicated that accumulation of this cellular subset depends on gut microflora. Importantly, in addition to $CD8\alpha$, other cytotoxic-related molecules, including GzmB, 2B4, and CRTAM, are induced in the absence of ThPOK. Consistent with these changes in gene expression, $CD4^+CD8\alpha\alpha^+$ cells acquire significant cytotoxic activity. These findings established that, in the context of gut-specific environmental cues, the functional capacity of naïve $CD4^+$ T cells can be reprogrammed to that of cytotoxic-like cells upon antigen stimulation. At a mechanistic level, the presence of a small population of ThPOK/CD8α double-negative cells, which are thought to be an intermediate stage before CD8α is reexpressed, indicates that ThPOK down-regulation precedes CD8α induction. A follow-up study extended this notion and showed that *Runx3* induction is important for reprogramming naïve CD4$^+$CD8$^-$ helper cells into this CD4$^+$ CD8$\alpha\alpha^+$ cytotoxic subset (Reis et al. 2013).

Together, these findings reveal a novel plasticity of CD4$^+$ T cells and challenge the dogma of stable commitment after helper versus cytotoxic-lineage decisions are made in the thymus. Although many recent studies have addressed developmental plasticity in CD4$^+$ T cells (Nakayamada et al. 2012), they do not clearly test whether mature CD4$^+$ T cells retain a plasticity for the acquisition of a functional cytotoxic phenotype in the periphery. However, armed with our current knowledge, literatures retrospective reveal that CD4$^+$ T cells bearing cytotoxic functions have long been identified (Appay et al. 2002; Brown 2010; Marshall and Swain 2011). The existence of this functional phenotype is now fully supported by the novel finding that mature CD4$^+$ T cells retain a flexibility to differentiate into cells with cytotoxic function in response to specific environmental cues. Reminiscent of lineage commitment in the thymus, the reciprocal expression of ThPOK and Runx3 is likely a key mechanism by which CD4$^+$ helper cells switch their functionality toward cytotoxic activity; a molecular process that is driven by Runx3-mediated engagement of the *Thpok* silencer (Mucida et al. 2013; Reis et al. 2013). Thus, the ThPOK/Runx3 axis is employed by immune cells not only during lineage commitment, but also to mediate flexibility in the functional differentiation of CD4$^+$ effector cells at mucosal borders.

6 Concluding Remarks

The identification of ThPOK and Runx3 as major regulators of the CD4/CD8 lineage decision has significantly advanced our understanding of how lineage segregation is controlled at the transcriptional level. The mechanisms that balance ThPOK and Runx3 expression are key features in the bifurcation of helper and cytotoxic lineages in the thymus. However, in addition to these central players, several other transcription factors contribute to the developmental programs that direct helper or cytotoxic-lineage commitment, presumably in cooperative and parallel manners with respect to ThPOK and Runx3. Surprisingly, recent studies have uncovered a previously unappreciated plasticity that enables naïve CD4$^+$ T cells to differentiate into CD4$^+$CD8$\alpha\alpha^+$ effector cells with cytotoxic function in the gut. Regulation of the ThPOK/Runx3 balance also serves as a key checkpoint for this extrathymic functional modification. Of note, induction of ThPOK in certain CD8$^+$ effector cells has been described as well, albeit at very limited amounts, and is potentially involved in regulating IL-2 production by memory CD8$^+$ T cells (Setoguchi et al. 2009). Thus, it is conceivable that an extrathymic balancing act between ThPOK and Runx3 expression is utilized to modify functions of both mature CD4- and CD8-lineage cells. Future studies addressing regulatory mechanism of ThPOK and Runx3 expression in the thymus and periphery will elucidate how environmental cues connect basic lineage programming in the thymus with functional plasticity in the periphery.

Acknowledgments This work was supported by grants from Grant-in-Aid for Scientific Research (S) and for Scientific Research on Priority Areas (I.T.). We are grateful to Dr. Hilde Cheroutre, Dr. Wooseok Seo, and Dr. Eugene Oltz for critically reading the manuscript and providing valuable advice.

References

Albu DI, Feng D, Bhattacharya D, Jenkins NA, Copeland NG, Liu P, Avram D (2007) BCL11B is required for positive selection and survival of double-positive thymocytes. J Exp Med 204:3003–3015

Aliahmad P, Kaye J (2008) Development of all CD4 T lineages requires nuclear factor TOX. J Exp Med 205:245–256

Aliahmad P, Kadavallore A, de la Torre B, Kappes D, Kaye J (2011) TOX is required for development of the CD4 T cell lineage gene program. J Immunol 187:5931–5940

Alvarez JD, Yasui DH, Niida H, Joh T, Loh DY, Kohwi-Shigematsu T (2000) The MAR-binding protein SATB1 orchestrates temporal and spatial expression of multiple genes during T-cell development. Genes Dev 14:521–535

Appay V, Zaunders JJ, Papagno L, Sutton J, Jaramillo A, Waters A, Easterbrook P, Grey P, Smith D, McMichael AJ, Cooper DA, Rowland-Jones SL, Kelleher AD (2002) Characterization of CD4$^+$ CTLs ex vivo. J Immunol 168:5954–5958

Bilic I, Koesters C, Unger B, Sekimata M, Hertweck A, Maschek R, Wilson CB, Ellmeier W (2006) Negative regulation of CD8 expression via Cd8 enhancer-mediated recruitment of the zinc finger protein MAZR. Nat Immunol 7:392–400

Brown DM (2010) Cytolytic CD4 cells: Direct mediators in infectious disease and malignancy. Cell Immunol 262:89–95

Brugnera E, Bhandoola A, Cibotti R, Yu Q, Guinter TI, Yamashita Y, Sharrow SO, Singer A (2000) Coreceptor reversal in the thymus: signaled CD4$^+$8$^+$ thymocytes initially terminate CD8 transcription even when differentiating into CD8$^+$ T cells. Immunity 13:59–71

Carpenter AC, Bosselut R (2010) Decision checkpoints in the thymus. Nat Immunol 11:666–673

Carpenter AC, Grainger JR, Xiong Y, Kanno Y, Chu HH, Wang L, Naik S, Dos Santos L, Wei L, Jenkins MK, O'Shea JJ, Belkaid Y, Bosselut R (2012) The Transcription Factors Thpok and LRF Are Necessary and Partly Redundant for T Helper Cell Differentiation. Immunity 37:622–633

Cruz-Guilloty F, Pipkin ME, Djuretic IM, Levanon D, Lotem J, Lichtenheld MG, Groner Y, Rao A (2009) Runx3 and T-box proteins cooperate to establish the transcriptional program of effector CTLs. J Exp Med 206:51–59

Dave VP, Allman D, Keefe R, Hardy RR, Kappes DJ (1998) HD mice: a novel mouse mutant with a specific defect in the generation of CD4$^+$ T cells. Proc Natl Acad Sci U S A 95:8187–8192

Djuretic IM, Levanon D, Negreanu V, Groner Y, Rao A, Ansel KM (2007) Transcription factors T-bet and Runx3 cooperate to activate Ifng and silence Il4 in T helper type 1 cells. Nat Immunol 8:145–153

Egawa T, Littman DR (2008) ThPOK acts late in specification of the helper T cell lineage and suppresses Runx-mediated commitment to the cytotoxic T cell lineage. Nat Immunol 9:1131–1139

Egawa T, Taniuchi I (2009) Antagonistic interplay between ThPOK and Runx in lineage choice of thymocytes. Blood Cells Mol Dis 43:27–29

Egawa T, Tillman RE, Naoe Y, Taniuchi I, Littman DR (2007) The role of the Runx transcription factors in thymocyte differentiation and in homeostasis of naive T cells. J Exp Med 204:1945–1957

The CD4/CD8 Lineages: Central Decisions and Peripheral Modifications

Ellmeier W, Sunshine MJ, Losos K, Hatam F, Littman DR (1997) An enhancer that directs lineage-specific expression of CD8 in positively selected thymocytes and mature T cells. Immunity 7:537–547

Ellmeier W, Sunshine MJ, Losos K, Littman DR (1998) Multiple developmental stage-specific enhancers regulate CD8 expression in developing thymocytes and in thymus-independent T cells. Immunity 9:485–496

Ellmeier W, Sawada S, Littman DR (1999) The regulation of CD4 and CD8 coreceptor gene expression during T cell development. Annu Rev Immunol 17:523–554

Gascoigne NR, Palmer E (2011) Signaling in thymic selection. Curr Opin Immunol 23:207–212

Germain RN (2002) T-cell development and the CD4-CD8 lineage decision. Nat Rev Immunol 2:309–322

Grenningloh R, Tai TS, Frahm N, Hongo TC, Chicoine AT, Brander C, Kaufmann DE, Ho IC (2011) Ets-1 maintains IL-7 receptor expression in peripheral T cells. J Immunol 186:969–976

Grusby MJ, Johnson RS, Papaioannou VE, Glimcher LH (1991) Depletion of CD4[+] T cells in major histocompatibility complex class II-deficient mice. Science 253:1417–1420

Hassan H, Sakaguchi S, Tenno M, Kopf A, Boucheron N, Carpenter AC, Egawa T, Taniuchi I, Ellmeier W (2011) Cd8 enhancer E8I and Runx factors regulate CD8alpha expression in activated CD8[+] T cells. Proc Natl Acad Sci U S A 108:18330–18335

He X, Dave VP, Zhang Y, Hua X, Nicolas E, Xu W, Roe BA, Kappes DJ (2005) The zinc finger transcription factor Th-POK regulates CD4 versus CD8 T-cell lineage commitment. Nature 433:826–833

He X, Park K, Wang H, He X, Zhang Y, Hua X, Li Y, Kappes DJ (2008) CD4-CD8 lineage commitment is regulated by a silencer element at the ThPOK transcription-factor locus. Immunity 28:346–358

Hedrick SM (2008) Thymus lineage commitment: a single switch. Immunity 28:297–299

Hosoya T, Maillard I, Engel JD (2010) From the cradle to the grave: activities of GATA-3 throughout T-cell development and differentiation. Immunol Rev 238:110–125

Hostert A, Garefalaki A, Mavria G, Tolaini M, Roderick K, Norton T, Mee PJ, Tybulewicz VL, Coles M, Kioussis D (1998) Hierarchical interactions of control elements determine CD8α gene expression in subsets of thymocytes and peripheral T cells. Immunity 9:497–508

Ikawa T, Hirose S, Masuda K, Kakugawa K, Satoh R, Shibano-Satoh A, Kominami R, Katsura Y, Kawamoto H (2010) An essential developmental checkpoint for production of the T cell lineage. Science 329:93–96

Jenkinson SR, Intlekofer AM, Sun G, Feigenbaum L, Reiner SL, Bosselut R (2007) Expression of the transcription factor cKrox in peripheral CD8 T cells reveals substantial postthymic plasticity in CD4-CD8 lineage differentiation. J Exp Med 204:267–272

Jones ME, Zhuang Y (2007) Acquisition of a functional T cell receptor during T lymphocyte development is enforced by HEB and E2A transcription factors. Immunity 27:860–870

Jones-Mason ME, Zhao X, Kappes D, Lasorella A, Iavarone A, Zhuang Y (2012) E protein transcription factors are required for the development of CD4[+] lineage T cells. Immunity 36:348–361

Kastner P, Chan S, Vogel WK, Zhang LJ, Topark-Ngarm A, Golonzhka O, Jost B, Le Gras S, Gross MK, Leid M (2010) Bcl11b represses a mature T-cell gene expression program in immature CD4[+]CD8[+] thymocytes. Eur J Immunol 40:2143–2154

Koller BH, Marrack P, Kappler JW, Smithies O (1990) Normal development of mice deficient in beta 2 M, MHC class I proteins, and CD8[+] T cells. Science 248:1227–1230

Konkel JE, Maruyama T, Carpenter AC, Xiong Y, Zamarron BF, Hall BE, Kulkarni AB, Zhang P, Bosselut R, Chen W (2011) Control of the development of CD8αα[+] intestinal intraepithelial lymphocytes by TGF-β. Nat Immunol 12:312–319

Lee SU, Maeda T (2012) POK/ZBTB proteins: an emerging family of proteins that regulate lymphoid development and function. Immunol Rev 247:107–119

Lee SU, Maeda M, Ishikawa Y, Li SM, Wilson A, Jubb AM, Sakurai N, Weng L, Fiorini E, Radtke F, Yan M, Macdonald HR, Chen CC, Maeda T (2012) LRF-mediated Dll4 repression in erythroblasts is necessary for hematopoietic stem cell maintenance. Blood 121:918–929

Li L, Leid M, Rothenberg EV (2010a) An early T cell lineage commitment checkpoint dependent on the transcription factor Bcl11b. Science 329:89–93

Li P, Burke S, Wang J, Chen X, Ortiz M, Lee SC, Lu D, Campos L, Goulding D, Ng BL, Dougan G, Huntly B, Gottgens B, Jenkins NA, Copeland NG, Colucci F, Liu P (2010b) Reprogramming of T cells to natural killer-like cells upon Bcl11b deletion. Science 329:85–89

Maeda T, Merghoub T, Hobbs RM, Dong L, Maeda M, Zakrzewski J, van den Brink MR, Zelent A, Shigematsu H, Akashi K, Teruya-Feldstein J, Cattoretti G, Pandolfi PP (2007) Regulation of B versus T lymphoid lineage fate decision by the proto-oncogene LRF. Science 316:860–866

Marshall NB, Swain SL (2011) Cytotoxic CD4 T cells in antiviral immunity. J Biomed Biotechnol 2011:954602

McCaughtry TM, Etzensperger R, Alag A, Tai X, Kurtulus S, Park JH, Grinberg A, Love P, Feigenbaum L, Erman B, Singer A (2012) Conditional deletion of cytokine receptor chains reveals that IL-7 and IL-15 specify CD8 cytotoxic lineage fate in the thymus. J Exp Med 209:2263–2276

Mucida D, Husain MM, Muroi S, van Wijk F, Shinnakasu R, Naoe Y, Reis B, Huang Y, Lambolez F, Docherty M, Attinger A, Shui JW, Kim G, Lena C, Sakaguchi S, Miyamoto C, Wang P, Atarashi K, Park Y, Nakayama T, Honda K, Ellmeier W, Kronenberg M, Taniuchi I, Cheroutre H (2013) Transcriptional reprogramming of mature CD4 T helper cells generates distinct MHC class II restricted cytotoxic T lymphocytes. Nat Immunol 14:281–289

Muroi S, Naoe Y, Miyamoto C, Akiyama K, Ikawa T, Masuda K, Kawamoto H, Taniuchi I (2008) Cascading suppression of transcriptional silencers by ThPOK seals helper T cell fate. Nat Immunol 9:1113–1121

Nakayamada S, Takahashi H, Kanno Y, O'Shea JJ (2012) Helper T cell diversity and plasticity. Curr Opin Immunol 24:297–302

Naoe Y, Setoguchi R, Akiyama K, Muroi S, Kuroda M, Hatam F, Littman DR, Taniuchi I (2007) Repression of interleukin-4 in T helper type 1 cells by Runx/Cbf beta binding to the Il4 silencer. J Exp Med 204:1749–1755

Park JH, Adoro S, Guinter T, Erman B, Alag AS, Catalfamo M, Kimura MY, Cui Y, Lucas PJ, Gress RE, Kubo M, Hennighausen L, Feigenbaum L, Singer A (2010) Signaling by intrathymic cytokines, not T cell antigen receptors, specifies CD8 lineage choice and promotes the differentiation of cytotoxic-lineage T cells. Nat Immunol 11:257–264

Pobezinsky LA, Angelov GS, Tai X, Jeurling S, Van Laethem F, Feigenbaum L, Park JH, Singer A (2012) Clonal deletion and the fate of autoreactive thymocytes that survive negative selection. Nat Immunol 13:569–578

Reis B, Rogoz A, Costa-Pinto F, Taniuchi I, Mucida D (2013) Mutual expression of Runx3 and ThPOK regulates intestinal CD4 T cell immunity. Nat Immunol 14:271–280

Rui J, Liu H, Zhu X, Cui Y, Liu X (2012) Epigenetic silencing of CD8 genes by ThPOK-mediated deacetylation during CD4 T cell differentiation. J Immunol 189:1380–1390

Sakaguchi S, Hombauer M, Bilic I, Naoe Y, Schebesta A, Taniuchi I, Ellmeier W (2010) The zinc-finger protein MAZR is part of the transcription factor network that controls the CD4 versus CD8 lineage fate of double-positive thymocytes. Nat Immunol 11:442–448

Sato T, Ohno S, Hayashi T, Sato C, Kohu K, Satake M, Habu S (2005) Dual functions of Runx proteins for reactivating CD8 and silencing CD4 at the commitment process into CD8 thymocytes. Immunity 22:317–328

Setoguchi R, Tachibana M, Naoe Y, Muroi S, Akiyama K, Tezuka C, Okuda T, Taniuchi I (2008) Repression of the transcription factor Th-POK by Runx complexes in cytotoxic T cell development. Science 319:822–825

Setoguchi R, Taniuchi I, Bevan MJ (2009) ThPOK derepression is required for robust CD8 T cell responses to viral infection. J Immunol 183:4467–4474

Singer A, Bosselut R (2004) CD4/CD8 coreceptors in thymocyte development, selection, and lineage commitment: analysis of the CD4/CD8 lineage decision. Adv Immunol 83:91–131

Singer A, Adoro S, Park JH (2008) Lineage fate and intense debate: myths, models and mechanisms of CD4- versus CD8-lineage choice. Nat Rev Immunol 8:788–801

Singh H (2007) Shaping a helper T cell identity. Nat Immunol 8:119–120

Sun G, Liu X, Mercado P, Jenkinson SR, Kypriotou M, Feigenbaum L, Galera P, Bosselut R (2005) The zinc finger protein cKrox directs CD4 lineage differentiation during intrathymic T cell positive selection. Nat Immunol 6:373–381

Tanaka H, Naito T, Muroi S, Seo W, Chihara R, Miyamoto C, Kominami R Taniuchi I (2013) Epigenetic Thpok silencing limits the time window to choose CD4+ helper-lineage fate in the thymus. EMBO J. 2013 Mar 12. doi:10.1038/emboj.2013.47

Taniuchi I, Osato M, Egawa T, Sunshine MJ, Bae SC, Komori T, Ito Y, Littman DR (2002) Differential requirements for Runx proteins in CD4 repression and epigenetic silencing during T lymphocyte development. Cell 111:621–633

Vanvalkenburgh J, Albu DI, Bapanpally C, Casanova S, Califano D, Jones DM, Ignatowicz L, Kawamoto S, Fagarasan S, Jenkins NA, Copeland NG, Liu P, Avram D (2011) Critical role of Bcl11b in suppressor function of T regulatory cells and prevention of inflammatory bowel disease. J Exp Med 208:2069–2081

Wang L, Wildt KF, Castro E, Xiong Y, Feigenbaum L, Tessarollo L, Bosselut R (2008a) The zinc finger transcription factor Zbtb7b represses CD8-lineage gene expression in peripheral CD4+ T cells. Immunity 29:876–887

Wang L, Wildt KF, Zhu J, Zhang X, Feigenbaum L, Tessarollo L, Paul WE, Fowlkes BJ, Bosselut R (2008b) Distinct functions for the transcription factors GATA-3 and ThPOK during intrathymic differentiation of CD4+ T cells. Nat Immunol 9:1122–1130

Wang Q, Stacy T, Miller JD, Lewis AF, Gu TL, Huang X, Bushweller JH, Bories JC, Alt FW, Ryan G, Liu PP, Wynshaw-Boris A, Binder M, Marin-Padilla M, Sharpe AH, Speck NA (1996) The CBFβ subunit is essential for CBFα2 (AML1) function in vivo. Cell 87:697–708

Woolf E, Xiao C, Fainaru O, Lotem J, Rosen D, Negreanu V, Bernstein Y, Goldenberg D, Brenner O, Berke G, Levanon D, Groner Y (2003) Runx3 and Runx1 are required for CD8 T cell development during thymopoiesis. Proc Natl Acad Sci U S A 100:7731–7736

Zamisch M, Tian L, Grenningloh R, Xiong Y, Wildt KF, Ehlers M, Ho IC, Bosselut R (2009) The transcription factor Ets1 is important for CD4 repression and Runx3 up-regulation during CD8 T cell differentiation in the thymus. J Exp Med 206:2685–2699

Zhang M, Zhang J, Rui J, Liu X (2010) p300-mediated acetylation stabilizes the Th-inducing POK factor. J Immunol 185:3960–3969

Zhu J, Min B, Hu-Li J, Watson CJ, Grinberg A, Wang Q, Killeen N, Urban JF Jr, Guo L, Paul WE (2004) Conditional deletion of Gata3 shows its essential function in T_H1–T_H2 responses. Nat Immunol 5:1157–1165

Zhu J, Yamane H, Cote-Sierra J, Guo L, Paul WE (2006) GATA-3 promotes Th2 responses through three different mechanisms: induction of Th2 cytokine production, selective growth of Th2 cells and inhibition of Th1 cell-specific factors. Cell Res 16:3–10

Index

A
Aire, 4, 22, 23, 25–32, 36, 38, 39, 74
APCs in Treg differentiation, 79
Autophagy, 8

C
Cathepsin L, 8
CCL21, 3
CCL25, 3
CCX-CKR1, 4
CD205, 3
Chemokines, 91, 94, 97–99, 102
Cortical thymic epithelial cells, 1, 53

D
DiGeorge syndrome, 2
DLL1, 5
DLL4, 5

F
Foxn1, 2

G
Gcm2, 2

H
Homeostatic proliferation, 58

I
IL-7, 7
Immunoproteasomes, 9

L
Lineage choice, 114, 115

M
Macroautophagy, 8
Medullary Epithelial Cells (mTECs), 72
MHC class I, 9
MHC class II, 8
Migration, 90, 91, 93, 94, 96, 97

N
Notch, 5
Nude mouse, 3

P
Plasticity, 123–125
Positive Selection, 7, 9, 50
Pre-TCR complex, 5
Progenitor, 88–102
Proteasomes, 9
Prss16, 8

S
Self-Peptides, 50

T
β5t, 4, 9
Tbx1, 2
T cell antigen receptor (TCR), 7
T cell development, 88, 95–97, 100
TGF-β, 7
Third pharyngeal pouch, 2

Thymic cortex, 2
Thymic DCs, 75
Thymic deletion, 71
Thymic involution, 4
Thymic nurse cell, 11
Thymocyte, 20–22, 24–28, 30–32, 35–39, 114–123
Thymoproteasome, 4, 9
Thymus, 20–31, 35–39, 88–90, 92–101

Thymus-specific serine protease, 8
T lymphocytes, 113, 121
Tolerance, 22, 25, 27–33, 39
Transcription factors, 115–121, 125
Treg niche, 77
Treg selection, 76
Treg self antigen specificity, 77
Treg TCR signal strength, 77